SpringerBriefs in Electrical and Computer Engineering

For further volumes:
http://www.springer.com/series/10059

Wen-Qin Wang

Near-Space Remote Sensing

Potential and Challenges

 Springer

Wen-Qin Wang
School of Communication and Information Engineering
University of Electronics Science
   and Technology of China (UESTC)
Chengdu
People's Republic of China
e-mail: wqwang@uestc.edu.cn

ISSN 2191-8112   e-ISSN 2191-8120
ISBN 978-3-642-22187-3   e-ISBN 978-3-642-22188-0
DOI 10.1007/978-3-642-22188-0
Springer Heidelberg Dordrecht London New York

Cover design: eStudio Calamar, Berlin/Figueres

Printed on acid-free paper

Springer is part of Springer Science+Business Media (www.springer.com)

Preface

Near-Space Remote Sensing: Potential and Challenges deals with the role of near-space vehicles in supplying a gap between satellites and airplanes for microwave remote sensing applications from a top-level system description, with an aim for further research.

Near-space is defined as the atmospheric region from about 20 to 100 km altitude above the Earth's surface. Very few sensors are currently operating in near-space, because the atmosphere is too thin to support flying for most aircrafts and yet too thick to sustain orbit for satellites. Nevertheless, potential benefits for vehicles operating in near-space may include possible persistent monitoring and high revisiting frequency (revisit the same site in a short time interval) that are critical to some specific radar and navigation applications, but not accessible for current satellites and airplanes.

There is a region in near-space where the average wind is <10 m/s; hence, persistent coverage and high flying speed can be obtained for the vehicles operating in this region. Moreover, near-space vehicles are relatively low cost when compared to satellites and airplanes. Additionally, as near-space is below ionosphere, therefore, there are no ionospheric scintillations that will significantly degrade microwave communication and navigation performance which explains why near-space has received much attention in recent years and why several types of near-space vehicles are being studied, developed, or employed.

Near-space vehicles offer the long-term persistence traditionally provided by satellites while providing the fast responsiveness of airplanes. Near-space thus offers an opportunity for developing new synthetic aperture radar (SAR) remote sensing techniques. Several potential applications such as passive surveillance, reconnaissance, high-resolution wide-swath remote sensing, and ground moving targets indication (GMTI) are described in this book.

Near-space vehicle-borne SARs cannot replace spaceborne and airborne radars, but they could provide more efficiently remote sensing functionality than spaceborne SARs. Although near-space vehicles have much smaller coverage area than satellites due to their lower altitude, they can still offer a regional coverage of hundreds of kilometers and provide cost-effective remote sensing services.

Near-space vehicle-borne SARs could also extend remote sensing services into areas with limited or no access to spaceborne and airborne SARs. Therefore, given their operational flexibility, near-space vehicle-borne SARs may supply a gap between spaceborne and airborne SARs which is the reason why we appeal to the system engineering community for more publication and support on the research and development of near-space vehicle-borne SARs.

With great pleasure I acknowledge the many people who have influenced my thinking and contributed to my knowledge. I express my deepest gratitude to Profs. Qicong Peng and Jingye Cai at the University of Electronic Science and Technology of China. They provided me with unprecedented freedom to spend my time on almost any topic that stimulated my curiosity. I also thank the support and encouragement of Prof. Xiaowen Li at the Institute of Remote Sensing Applications, Chinese Academy of Sciences.

I also wish to acknowledge the support provided by the Doctoral Program of Higher Education for New Teachers under contract *200806141101*, Fundamental Research Funds for the Central Universities under contract *ZYGX2010J001*, and the Open Funds of the State Laboratory of Remote Sensing Science under contract *OFSLRSS201011*, the National Key Laboratory of Millimeterwave Technology under contract *K200914* and the Key Laboratory of Ocean Circulation and Waves, Chinese Academy of Sciences under contract *KLOCAW1004*.

As always the support of my wife Ke Yang is gratefully acknowledged for her constant and gentle encouragement.

Finally, I thank Ms. Becky Zhao and Na Xu from Springer for their wonderful help in the preparation and publication of this manuscript.

People's Republic of China, June 2011   Wen-Qin Wang

Contents

1 Introduction 1
  1.1 Background 1
    1.1.1 What is Near-Space... 1
    1.1.2 Near-Space Environment 2
    1.1.3 Why Near-Space Remote Sensing................................ 3
  1.2 Outline of the Chapter 3
  References 4

2 Near-Space Vehicles: Remote Sensing Advantages................... 5
  2.1 Near-Space Vehicles 5
    2.1.1 Free-Floaters 5
    2.1.2 Steered Free-Floaters.. 6
    2.1.3 Maneuvering Vehicles....................................... 7
  2.2 State-of-the-Art.. 8
    2.2.1 In North America 8
    2.2.2 In Europe.. 9
    2.2.3 In Asia-Pacific 10
  2.3 Comparative Advantages 11
    2.3.1 Inherent Survivability 11
    2.3.2 Persistent Region Coverage or High-Revisiting
        Frequency 12
    2.3.3 Relative High Sensitivity and Large Footprint 13
    2.3.4 Low Cost.. 13
  2.4 Limitations and Vulnerabilities 14
    2.4.1 Launch Constraints 14
    2.4.2 Survivability Constraints 14
    2.4.3 Legal Constraints 15
  2.5 Concluding Remarks 15
  References 16

3   Near-Space Vehicles in Passive Remote Sensing   19
   3.1   Near-Space Vehicles in Passive Surveillance   19
      3.1.1   System Configuration   19
      3.1.2   Signal Models   20
      3.1.3   Target Location   22
      3.1.4   Power Budget Analysis   23
   3.2   Near-Space Vehicles in Passive BiSAR Imaging............   24
      3.2.1   System Imaging Performance....................   25
      3.2.2   Azimuth-Variant Characteristics   30
      3.2.3   Two-Dimensional Spectrum Model   31
      3.2.4   Image Formation Processing.....................   33
   3.3   Near-Space Vehicles in Passive Environment Monitoring......   36
   3.4   Potential and Challenges   38
      3.4.1   Potential Applications: Homeland Security...........   39
      3.4.2   Potential Applications: Persistently
              Disaster Monitoring..........................   39
      3.4.3   Challenges: Synchronization Compensation   40
      3.4.4   Challenges: Motion Compensation   43
      3.4.5   Challenges: Antenna Directing Synchronization   45
   3.5   Conclusion   47
   References   47

**4   Near-Space Vehicles in High-Resolution Wide-Swath
     Remote Sensing**   51
   4.1   Restrictions on Achievable Resolution and Swath   51
   4.2   State-of-the-Art: HRWS Remote Sensing   53
      4.2.1   Multiple Apertures in Elevation   53
      4.2.2   Multiple Channels in Azimuth   54
      4.2.3   Multiple Apertures in Two Dimensions   55
      4.2.4   Distributed SAR Constellations...................   55
   4.3   Near-Space Vehicle-Borne SAR HRWS Remote Sensing......   56
      4.3.1   Single-Phase Center Multibeam SAR Imaging   56
      4.3.2   Multiple Phase Center Multibeam SAR Imaging.......   59
      4.3.3   Ambiguity-to-Signal Ratio Analysis   64
      4.3.4   Conceptual System Design......................   67
   4.4   Near-Space HRWS Remote Sensing via Multiple Apertures   67
      4.4.1   System Architecture and Imaging Scheme   68
      4.4.2   Imaging Performance Analysis   72
      4.4.3   Conceptual Examples and Simulation Results.   74
   4.5   Near-Space HRWS Remote Sensing via Waveform Diversity ...   77
      4.5.1   Waveform Diversity Design.....................   78
      4.5.2   MIMO SAR-Based Wide-Swath Remote Sensing   81

4.5.3   Space-Time Coding MIMO SAR for High-Resolution
          Imaging   85
4.6   Conclusion   94
References   95

5 **Near-Space Vehicles in Ground Moving Target Indication**   99
5.1   MIMO SAR with Multi-Antenna in Azimuth   99
5.2   MIMO SAR-Based GMTI Processing   101
5.3   Simplified FrFT-Based Parameters Estimation   105
       5.3.1   Simplified FrFT Algorithm   105
       5.3.2   Simplified FrFT-Based Estimation Algorithm.   106
5.4   Simulation Results.   108
5.5   Conclusion   109
References   109

6 **Summary**.   111
6.1   Realistic Near-Space Remote Sensing Issues.   111
6.2   Future Work   112
       6.2.1   High-Precision Imaging Algorithm   112
       6.2.2   Waveform Diversity Design.   112
       6.2.3   Three-Dimensional Imaging   113
References   114

Chapter 1
Introduction

Abstract Near-space, the region between controlled commercial airspace and low-earth orbit (LEO), offers several capabilities that are critical to emerging remote sensing applications, but not accessible to current satellites and airplanes. In this chapter, we explain what near-space is and why it should be exploited for remote sensing.

Keywords Near-space · Space definition · Near-space environment · Satellite · Airplane · Remote sensing

1.1 Background

Near-space was an area out of reach of airborne aircraft, yet below the effective satellite orbit. However, in recent years new technology and interest have brought about a reexamining of this place between air and space [1].

1.1.1 What is Near-Space

Near-space is defined as the atmospheric region from about 20 km altitude to 100 km altitude above the Earth's surface [2], as shown in Fig. 1.1. Note that the lower limit is not determined from operational considerations, but from the international controlled airspace altitude [3].

Traditionally very few sensors are currently operating in near-space because the atmosphere is too thin to support flying for airplanes and yet too thick to sustain orbit for satellites. However, evolutionary advances in several technologies have a revolutionary advance in capability. One technology is the power supplies including thin, light-weight solar cells, small, efficient fuel cells, and high-energy-density batteries; the miniaturization of electronics and exponential increase in computing power, enabling extremely capable sensors in very small and light-weight packages;

W.-Q. Wang, *Near-Space Remote Sensing*,
SpringerBriefs in Electrical and Computer Engineering,
DOI: 10.1007/978-3-642-22188-0_1, © Wen-Qin Wang 2011

Fig. 1.1 Near-space
definition and its advantages
while compared to space
including geosynchronous
orbit (*GEO*), middle earth
orbit (*MEO*) and *LEO*, and
airplane

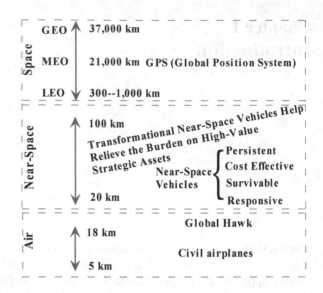

and very light-weight, strong, and flexible materials that can resist degradation under strong ultraviolet illumination and relatively impermeable to helium or hydrogen [4].

Another two emerging technologies are high-altitude buoyant lift systems and plasma thrust technology [2]. As altitude increases, electrodynamic power transfers into the air which can be used for propulsion, cooling, and control. This allows us to call upon many creative electromagnetic circuits for our designs. Plasma technology supports the high-altitude and prompt global strike programs pursued by many institutes or laboratories.

Near-space thus has become an area of exceptional interest in the last several years. Possible real-world uses in communication, radar, and navigation applications have come to light.

1.1.2 Near-Space Environment

There are no clouds, thunderstorms, or precipitation in near-space. Although the air density in near-space is very low, wind is still an important environment factor. Wind in near-space varies with altitude, time of year, and latitude, generally increasing with both latitude and altitude. In higher near-space there is no appreciable wind. Note that sustained winds will have an effect, but guests are more important to be considered when designing the near-space vehicles that have requirements to maneuver.

Another important advantage is that near-space vehicles operate below ionosphere. Ionosphere is a region containing several distinct layers of charged particles surrounding the earth. Its main layers occur at about 70–90, 95–140, and 200–400 km above the earth. Ionosphere affects the electromagnetic signals that pass through it.

Depending on frequency and direction of propagation, some signals are slowed, attenuated, or bent slightly while others can be completely absorbed or bent. Spacecraft thus needs to be designed to mitigate operations in this highly-charged environment while near-space vehicles can avoid this cost. Therefore, in near-space there are no ionospheric scintillations that will significantly degrade microwave communication and navigation performance which explains why near-space has received much attention in recent years and why several types of near-space vehicles are being studied, developed, or employed [5, 6].

1.1.3 Why Near-Space Remote Sensing

Current spaceborne and airborne synthetic aperture radars (SARs) are placing an important role in remote sensing applications; however, even as good as they envisioned or employed, it is impossible for our limited non-geosynchronous earth orbit satellites to provide a staring presence on a timescale of days, weeks, or months over a selected target or area of interest. Even if we can launch a satellite for particular application, it would only be in view for a short time. As an example, most LEO satellites have a specific target in view for less than 10 min at a time. In contrast, conventional airplanes cannot fly too high because there is no sufficient oxygen to allow the engines to operate. Generally, satellites operate in the orbits above 200 km and air-breathing airplanes cannot routinely operate much higher than 18 km.

Thus we have two gaps. The first is a gap in capability of persistent remote sensing observation. The second is a gap where there is little sensors in the altitude between air and space. To overcome these disadvantages, some alternative platforms should be employed [7]. Moreover, the requirements of stealth and robust survivability in military applications also call for new radar platforms other than satellites and airplanes. Fortunately, these aims can be simultaneously obtained by using near-space platforms, at a fractional cost of the traditional platforms [8].

In fact, by placing SAR transmitters and/or receivers inside near-space vehicles, many functionalities that are currently performed with satellites or airplanes could be performed more efficiently than conventional airborne and spaceborne SARs. Some works on the use of near-space vehicle-borne sensors for communication applications have been reported [9]. Some other works have been reported on near-space vehicles in radar and navigation applications [3, 10–13].

1.2 Outline of the Chapter

In this book, we discussed the role of near-space vehicles in supplying a gap between satellites and airplanes for remote sensing applications from a top-level system description, with an aim for further research. Several potential applications and corresponding challenges are described. The remaining chapters are organized as follows.

Chapter 2 investigates the advantages of near-space vehicles in remote sensing, while compared to satellites and airplanes. Near-space vehicles can be classified into free-floater, steered free-floater, and maneuvering vehicle. Each category is further explained in great detail.

Chapter 3 is devoted to near-space vehicles in passive microwave remote sensing. Applications of this technique to regional passive remote sensing and challenges of synchronization and motion compensations are given as examples.

Chapter 4 deals with near-space vehicles in wide-swath remote sensing. The constraints on high-resolution and wide-swath in SAR remote sensing are derived. Several multi-antenna or multi-aperture techniques, such as multi-channel in azimuth and multi-channel in elevation, are proposed.

Chapter 5 proposes a scheme of near-space vehicle-borne multiple-input and multiple-output (MIMO) SAR for ground moving targets indication (GMTI) applications. This approach is more effective and robust than the conventional displaced phase center antenna (DPCA) SAR-based GMTI solutions.

Chapter 6 provides a concise summary of the near-space remote sensing and discusses the realistic issues and future work.

References

1. Tomme, E.B.: The paradigm shift of effects-based space: near-space as a combat space effects enabler. http://www.airpower.au.af.mi (2009). Accessed May 2010
2. Allen, E.H.: The case for near-space. Aerosp. Am. **22**, 31–34 (2006)
3. Wang, W.Q., Cai, J.Y., Peng, Q.C.: Near-space microwave radar remote sensing: potentials and challenge analysis. Remote Sens. **2**, 717–739 (2010)
4. Tomme, M., Dahl, Z.: Balloons in today's military: an introduction to the near-space concept. Air Space Power J. **2**, 39–50 (2005)
5. Marcel, M.J., Baker, J.: Interdisciplinary design of a near-space vehicle. In: Proceedings of Southeast Conference, pp. 421–426. Richmond, VA (2007)
6. Romeo, G., Frulla, G.: HELIPLAT: high altitude very-long endurance solar powered UAV for telecommunication and earth observation applications. Aeronautical J. **108**, 277–293 (2004)
7. Wang, W.Q., Cai, J.Y.: A technique for jamming bi- and multistatic SAR systems. IEEE Geosci. Remote Sens. Lett. **4**, 80–82 (2007)
8. Wang, W.Q.: Application of near-space passive radar for homeland security. Sens. Imag. Int. J. **8**, 39–52 (2007)
9. Guan, M.X., Guo, Q., Li, L.: A novel access protocol for communication system in near-space. In: Proceedings of Wireless Communication and Network Mobile Computation Conference, pp. 1849–1852. Shanghai, China (2007)
10. Wang, W.Q., Cai, J.Y., Peng, Q.C.: Near-space SAR: a revolutionay microwave remote sensing mission. In: Proceedings of Asia-Pacific Synthetic Aperture Radar Conference, pp. 127–131. Huangshan, China (2007)
11. Galletti, M., Krieger, G., Thomas, B., Marquart, M., Johannes, S.S.: Concept design of a near-space radar for tsunami detection. In: Proceedings of IEEE Geoscience and Remote Sensing Symposium, pp. 34–37. Barcelona (2007)
12. Wang, S.Y., Tao, C., Chen, D.: Research on guidance under multiple constraints for near space vehicles. In: Proceedings of System Control in Aeronautics Astronautics Symposium, pp. 1261–1264. Harbin, China (2010)
13. Wang, W.Q., Cai, J.Y., Peng, Q.C.: Passive ocean remote sensing by near-space vehicle-borne GPS receiver. In: Tang, D.L. (ed.) Remote Sensing of the Changing Oceans. Springer-Verlag, Berlin (2011)

Chapter 2
Near-Space Vehicles: Remote Sensing Advantages

Abstract Near-space provides a promise for future remote sensing applications. Instead of concentrating payloads, in this chapter we give a brief overview of the basic types of near-space vehicles currently in use, in active development, or envisioned. Their advantages, limitations, and vulnerabilities for microwave remote sensing are investigated.

Keywords Near-space · Remote sensing · Near-space vehicle · Free-floater · Maneuvering vehicle · Persistent coverage

2.1 Near-Space Vehicles

Some of the near-space vehicles already exist and a great deal of new near-space vehicles are currently in prototype [1, 2]. Near-space vehicles can be classified into three major categories [3]: (1) free-floaters, (2) steered free-floaters, and (3) maneuvering vehicles. Free-floaters are like rudimentary rafts where the speed and direction of travel are completely determined by the direction of the current. Steered free-floaters are like sailboats where the current has a large effect on their motion but they can steer within that current. Maneuvering vehicles are akin to steamships: they can go where they want and stay there for as long as they like.

2.1.1 Free-Floaters

Free-floaters are essentially large balloons that float with the wind. They are normally manufactured in two types: zero-pressure and super-pressure. Zero-pressure free-floaters are similar to weather and recreational balloons, which have a venting system which ensures that the pressure inside the balloon is the same as the surrounding atmosphere. They are less vulnerable to puncture, since significant amount of the lifting gas must diffuse out before the list is lost. Imaging an inflated, light-weight

W.-Q. Wang, *Near-Space Remote Sensing*,
SpringerBriefs in Electrical and Computer Engineering,
DOI: 10.1007/978-3-642-22188-0_2, © Wen-Qin Wang 2011

Fig. 2.1 Example
free-floater. Image courtesy
USA Air Force Space Battle
Lab

plastic garment bag floating on the wind; even if there are many holes in such a bag, it still can float in the air for a long time. In contrast, super-pressure free-floaters have a higher pressure inside than outside. They are constructed by strong materials, making them relatively insensitive to puncture damage.

Once launched, free-floaters are at the mercy of the existing winds. Limited steering is possible by variable ballasting, causing them to float at different altitudes to take advantage of different wind directions and speeds. However, they have no station-keep ability because no active steering or propulsion techniques are employed in these platforms. Free-floaters have already demonstrated commercial viability as communication platforms, e.g. Fig. 2.1. They can lift payloads of tens to thousands of pounds to over 30 km, depending on their volume [4].

The biggest drawback of most free-floaters is their payloads generally cannot be recovered. In recent years, by encasing the payloads in a high-performance autonomous glider, expensive or sensitive payloads can be recovered safely and reused. When the floater approaches the maximum range of the glider, the glider is cut loose from the floater. The payload then autonomously glides back from hundreds of kilometers away, staying aloft for several hours before landing safely on ground. A variety of such gliders are available today, ranging from extremely inexpensive plastic gliders with limited payload capability to much more complex and capable composite gliders.

2.1.2 Steered Free-Floaters

Steered free-floaters also drift on the wind, but they are able to exploit the wind much like sailing ships to maneuvering almost at will. Sailing requires the vehicle to be immersed in two media moving at different speeds. A large balloon at high altitude moves at a different speed through the air than a wing suspended below the balloon at a different altitude. The air around the wing is moving at a different speed than the air pushing the balloon. The entire platform is then steered when the differential wind between the two parts of the platform enables the wing to become aerodynamically

Fig. 2.2 Example steered
free-floater. Image courtesy
USA Air Force Space Battle
Lab

Fig. 2.3 Example
maneuvering vehicles.
Image courtesy USA Naval
Research Laboratory

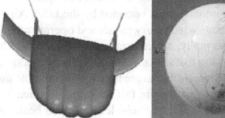

effective. Figure 2.2 gives an example of steered free-floater of the USA Air Force
Space Battle Lab. Steered free-floater technology has been commercially mature and
military deployment is imminent.

Steered free-floaters can be navigated with fairly high degree of precision, generally going with the flow of the prevailing latitudinal winds and being able to speed
up, slow down, and move perpendicular to those winds. With the limited steering,
these balloons can stay on station for a short time. Their payloads could be more
complex than those flown on basic free-floaters.

2.1.3 Maneuvering Vehicles

Maneuvering vehicles has a means of propulsion and a means of control. The propulsion can rely on fossil fuel, nuclear, or solar energy. The control can be attained
through both aerodynamic and aerostatic means. They can maneuver and thus fly to
and station-keep over a specific area of interest. Maneuvering vehicles are the functional cross between satellites and airplanes. They can provide large footprint and
long-mission durations that are commonly associated with satellites and responsiveness that is commonly associated with unmanned aerial vehicles. Figure 2.3 gives
one conceptual version of the USA navy's high altitude airborne relay and router.

Fig. 2.4 Example tethered
balloon-like aerostat

Maneuvering vehicles can bring large payloads (up to 1,000 tons) that are enough
to be useful for remote sensing applications and can be recovered for repair and reuse.
Persistent coverage is also possible for maneuvering vehicles. One tethered balloon-
borne radar system (e.g., Fig. 2.4) has been operationally used in an air defense and
drug enforcement network operated by the USA Air Force. This network uses two
sizes of aerostats which carry radars and operate at an altitude of 4.57 km feet.

More importantly, not constrained by orbital mechanics like satellites and high-
fuel consumption like airplanes, they can move at a speed as fast as 1,000–1,500 m/s.
Thus, maneuvering vehicles are potentially the most useful type for the remote sensing applications that require fast revisiting frequency and high-resolution imaging
performance. Maneuvering vehicles have also been viewed as a possible substitute
for satellites supporting communications and other missions ([5–19]).

2.2 State-of-the-Art

Recently many projects focusing on near-space vehicles have been funded. In this
section, the major near-space projects and players are briefly discussed. We focus
here on their major achievements, including trials and demonstrations.

2.2.1 In North America

Sky Station was a North American commercial project consisting of a solar-powered
aerostatic high-altitude platform system planned by Sky Station International [20].
This platform could be maintained geo-stationary at altitudes of approximately 21 km
in the stratosphere, over major metropolitan areas. The average platform dimension
could be 200 m long and 60 m in diameter at its widest point. Sky Station International
planned to deploy at least 250 platforms, one about 21 km above every major city
in the world for wireless communication services. An overview of the Sky Station
project can be found in Ref. [21].

Another North American commercial project is the Pathfinder, Pathfinder plus,
HELIOS, and SkyTower managed by USA AeroVironment Corporation. The first

Pathfinder was designed and fabricated in the early 1980s to support a classified program. It concluded that the required technology had not reached a level where ultra long-duration flight under solar power could be achieved. In 1998, Pathfinder was modified by NASA into the longer-winged Pathfinder Plus for civil telecommunications. This aircraft achieved the record of a flight altitude of 24 km. The following HELIOS was capable of continuous flight for up to 6 months or more at 21 km [22]. The SkyTower aims to market high-altitude platforms to deliver a wide range of applications, such as fixed broadband communications, narrowband and broadcast communications. The last of the key technology work was completed in 2005, and practical flight and ground operations were demonstrated in 2006 [23].

The primary ongoing North American military effort to develop near-space vehicles is the advanced concept technology demonstration (ACTD), which was initiated in 2003 to design, build, and test a high-altitude aerostat prototype that is able to operate unmanned, maintain a geostationary position over 21 km for up to 6 months, generate its own power, and carry a variety of payloads. In 2003 Lockheed Martin Naval Electronics and Surveillance Systems was selected by the U.S. Missile Defense Agency to perform the ACTD Phase II contract award. The $40,000,000 contract through June 2004 calls for the design of a solar-powered high-altitude airship. The airship is planned to have a mission life of one month, operating at 20 km, while providing 10 kW of power to a 4,000 pound payload. It is intended that it will become a part of the Ballistic Missile Defense System Test bed. Some of the early activities demonstrated the capabilities of station-keeping and autonomous flight. It will be used for military and civilian activities including: (1) weather and environment monitoring, (2) short- and long-range missile warning, (3) surveillance, and (4) target acquisition. The ACTD program documentation summarizes the effort as having "some technical risk" but "enormous potential benefits".

2.2.2 In Europe

In Europe, mainly two organizations have funded research activities throughout the continent, the European Space Agency (ESA) and the European Commission (EC). The ESA was one of the first research institutions to promote research on near-space vehicles in the continent. The EC has funded projects to develop and make demonstrators for near-space vehicles and to promote strategic studies related to the future of this technology. The representative research projects include HeliNet (network of stratospheric platforms for traffic monitoring, environmental surveillance, and broadband services), CAPANINA (broadband communications technology), UAVNET (unmanned air vehicles network), CAPECON (civil UAV applications and economic affectivity of potential configuration solutions), and USICO (UAV safety issues for civil operations).

In 1998 the ESA awarded to the Lindstrand Balloons Ltd a design contract for a geostationary stratospheric unmanned airship. This airship should fly in the stratosphere at 21 km altitude and could remain stationary over its intended position [24]. The airship was designed to carry a 600 kg payload with a relay station, surveillance radar, and a weather radar or sensor package. It mainly devoted to the provisions of civil services in synergy or competition with both satellite and terrestrial systems. The STRATOS was another ESA-funded research project. The main object was to explore the grounds for the development and operation of a European near-space vehicle system based on the performance of a conceptual design for the best suited stratospheric platform concept answering the needs of future telecommunication markets [25].

The HeliNet was a project based upon high altitude very long endurance unmanned solar aerodynamic platforms, funded by the 5th Framework Programme of the European Union Commission. HeliNet was a global project carried out by a transnational and multi-sectorial partnership of research departments at universities and companies from Italy, Spain, UK, Slovenia, Hungary, and Switzerland. The prototype was based on the design of an unmanned solar-powered aircraft, named Heliplat. Heliplat was tailored for long-endurance operations at an altitude of 17 km, supporting a payload of 100 kg and offering a power of 800 W. Apart from the Heliplat, three prototype applications were examined as well, i.e., broadband telecommunications, remote sensing, and navigation/location [26, 27]. After the successful completion of the HeliNet project in 2003, the EC started the CAPANINA research project, which is being partially funded by the 6th European Union's Framework initiative. Built on the HeliNet project, CAPANINA aims at the development of low-cost broadband technology from near-space vehicles to deliver cost-effective solutions to users in urban and remote rural areas, or to users traveling inside high-speed public transport vehicles at speeds up to 300 km/h. One of the greatest achievements of CAPANINA was the fact that it proved credibility in the use of near-space vehicles to deliver broadband services. Three trials were completed successfully. The first trial was conducted in UK in 2004 using a very low-altitude tethered airship platform at 300 m altitude. The second trial was conducted and completed in Sweden in 2005 using a free-floater that could reach 25 km altitude. The last trial was conducted in USA in 2007 using a free-flight stratospheric balloon.

2.2.3 In Asia-Pacific

Asia-Pacific projects and activities on near-space vehicles were mainly undertaken by Japan, Korea, China, Australia, New Zealand, and Malaysia. Skynet was one Japanese project for the development of a balloon based on stratospheric platforms capable of operating at an altitude of 20 km and carrying on-board mission payloads for communications, broadcasting, and environmental observations. In the framework of this project, two prototype airships have been developed. The first airship, named Ground-to-stratosphere, has no propulsion system, and was successfully used to obtain thermal, buoyancy, and position control. It can ascend to the altitude of 15 km and descend to a planned area in the ocean. The second airship, named low-altitude-stationary, has a propulsion system. It will be used in order to obtain station-keeping mechanisms. The propulsive propellers were mounted on both the stem and stern of the airship. A solar power subsystem of solar cells and regenerative fuel cells was provided to supply a day-night cycle of electricity for airship propulsion [28].

In Korea research activities on near-space vehicles were conducted mainly by ETRI (Electronics and Telecommunications Research Institute), whereas KARI (Korean Aerospace Research Institute) deals with airship research and development [29]. The main aim is to develop an unmanned stratospheric airship and ground systems for basic operation and control of the airship. This research program consists of three phases. In the first phase one 50 m scale-size unmanned airship was built. The second phase aims to demonstrate the feasibility of developing a stratospheric platform that could be used at around 20 km altitude. The third phase aims to develop a full-scale 200 m airship that could carry telecommunications and remote sensing payloads weight up to 1,000 kg.

In 2007 international agreements were achieved among the Malaysian government, QucomHaps Co. Ireland and the proprietor of Russian M-55GN stratospheric aircraft. The main aim was to provide cost-effective nation-wide wireless access to broadband connectivity using M-55GN stratospheric aircraft. The M-55GN is a piloted plane, which can fly in a circular corridor at an altitude of approximately 21 km with a flight endurance of approximately 5 h. It is an all-weather single-seater stratospheric aircraft capable of operating both day and night, even in critical environmental conditions and strong cross-winds. Additionally, Chinese National Natural Science Foundation have also funded several research projects on near-space vehicles [30–33].

2.3 Comparative Advantages

While compared to current satellite and airplane platforms, near-space vehicle platforms have many superiorities for microwave remote sensing applications [34]:

2.3.1 Inherent Survivability

Near-space vehicles, especially free-floaters, have an inherent survivability. Free-floaters have extremely small radar cross and thermal cross-sections, making them relatively invulnerable to most traditional tracking and positioning methods. Estimates of free-floater's radar cross-section (RCS) are on the order of hundredths of a square meter. In fact, at near-space altitude, free-floaters will be small optical targets, only showing up well when the background is much darker than them. Consequently, the acquisition and tracking problem will be very difficult, even without considering what sort of weapon could reach them. Surface-to-air missiles (SAMs) may be a threat, but they are most likely not designed to engage a non-maneuvering target at that altitude. Economics also discourage such an exchange because the trade between

Table 2.1 The maximum observation time (min:s) for selected LEO satellites at different orbital altitude (km) and different angle (degrees) above horizon

Orbital altitude (km)	0°	5°	10°	30°	45°
200	7:49	5:37	4:08	1:40	1:00
300	9:35	7:16	5:34	2:24	1:27
400	11:10	8:44	6:54	3:08	1:54

an inexpensive near-space vehicle and a missile is cost-prohibitive. In modern electronic battlefield, enemies may only want to launch their missiles for higher value and immediately destroyable targets. Free-floaters floating across their territory in numbers may not be worth of the effort and expense to shoot down. Moreover, even if the acquisition and tracking problems are overcome, near-space vehicles are difficult to destroy.

On the other hand, there are defensive options available to help near-space vehicle's survivability. One such option could be deceptive "chaff" similar to that carried on modern fighter aircraft. This chaff could be dispensed in hopes of confusing radar guidance. Also, using vehicle decoys is an option. The vehicles themselves are relatively inexpensive and a simple fake payload could be attached. They might have imitated electronic and infrared signatures, further making it difficult to discriminate between the vehicles.

2.3.2 Persistent Region Coverage or High-Revisiting Frequency

The most useful and unique aspect of near-space vehicles is their ability to provide persistent region coverage or high-revisiting frequency. Space technologies have significantly revolutionized modern battlefield and remote sensing; however, persistent coverage, which is highly desired, is still unavailable through satellites or airplanes. Table 2.1 shows the observation time for selected LEO orbits. Air-breathing airplanes provide responsive persistence for the duration of their limited loiter times. The longest persistence that we can currently expect from an air-breathing asset is about a day or so.

Fortunately, persistent coverage can be achieved by using near-space vehicles. The altitude of near-space is above troposphere and atmosphere region where most weather occurs. There is no cloud, thunderstorm, or precipitation. Moreover, propulsion technique can be further applied to countermine possible mild winds in near-space. Near-space free-floaters can thus stay at a specific site for a long time, a persistence is thus possible. This advantage has a particular value for the applications that require persistent monitoring.

Near-space maneuvering vehicles can use a variety of schemes for propulsion, including conventional propellers and unconventional buoyancy-modification schemes that allow the vehicles to propel themselves. Maneuvering vehicles are the functional cross between satellites and airplanes. They can fast-fly or station-keep over a specific position, to provide large footprint and long-mission duration.

Fig. 2.5 Ground coverage area as a function of looking-down angle for different flying altitude

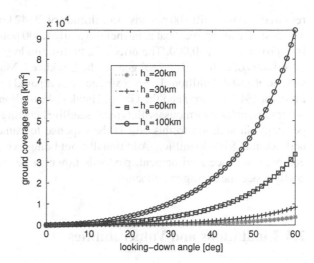

The near-space vehicles currently receiving technology demonstration funding will be able to stay on station for 6 months, and planned follows are projected to stay aloft for years [4].

2.3.3 Relative High Sensitivity and Large Footprint

Near-space vehicles are much closer to the targets than their orbital cousins. Distance is critical to receiving low-power signals. From the radar equation [35] we know that the received signal power attenuates at a square of the distance from the transmitter to the target, while that of an active antenna attenuates at a fourth power of the transmitter distance. Considering a point at nadir, near-space vehicles are 10–20 times closer to their targets than a typical 400 km LEO satellite. This distance differential implies that it could detect much weaker signals (10–13 dB weaker).

On the other hand, near-space vehicles will have an impressive footprint from such high altitudes. Figure 2.5 shows the ground coverage area as a function of looking-down angle for different flying altitude. Although orbiting satellites traditionally have a much larger footprint than near-space vehicles, they do so at the cost of persistence and signal strength. Additionally, being lower than satellites also brings about another advantage to near-space vehicles: they fly below the ionosphere. Ionospheric scintillation is very difficult to predict, but it can disrupt satellite communication and navigation performance significantly. Fortunately, ionospheric scintillation has no impact on near-space vehicles.

2.3.4 Low Cost

The inherent simplicity, recoverability, and relative less requirements of complex infrastructure all contribute to this advantage. The cost of a free-floater and the required helium to lift 100 pounds to an altitude of 20.42 km is approximately $1,400. The cost of a larger free-floater and helium to lift 1,000 pounds up to the same altitude is approximately $10,000. The price of a high-altitude vehicle will be on the order of millions of dollars. The final goal of the Lockheed–Martin is to keep the cost per vehicle at $50–60 million. These costs can be compared to current UAV's such as the Predator ($4.5 million) and the Global Hawk ($48 million). Also, satellite expenses are traditionally enormous, with typical satellites costing roughly $60–300 million per copy. In addition to this cost is the expense to launch them into orbit, which adds another $10–40 million. Additionally, not being exposed to electronic radiation common to space environment, payloads flown in near-space vehicles require no costly space-hardening manufacture.

2.4 Limitations and Vulnerabilities

Near-space vehicles also have several limitations and vulnerabilities. The most prominent weaknesses are launch constraints, survivability constraints, and legal constraints.

2.4.1 Launch Constraints

Weather will be a risk factor that could be significant if vehicles are not furnished with reliable sensors for on-site meteorological data with which vehicle controllers can predict turbulence, icing, and violent gusts that jeopardize the craft. The experience with high-altitude tropospheric operation from around-the-world balloonist teams and weather teams must be collected and codified to aid computer predictions at higher altitudes. As a vehicle will be in the troposphere for over five hours while descending to its home-mooring base, the weather conditions will have to be within allowable parameters before a letdown can commence. This requirement could cause the vehicle to hold at near-space for up to two–five days before descending. Launch operations could cause similar delays.

Note that satellites face similar launch constraints, but those constraints only have to be met during launch. UAVs and manned aircraft are also subject to similar launch and recovery constraints, although their limitations are less stringent than those for near-space vehicles.

2.4.2 Survivability Constraints

It is a hasty assumption to think that near-space vehicles will continue to remain out of harm's way in the future. As near-space technology develops, it should be assumed that enemy technology will also develop, and air defense systems will improve. Near-space was not historically a viable region to operate and missiles were not developed to reach that high. However, a modern surface-to-air missile (e.g., SA-2) can reach as high as jet aircraft flew. The shoot-down of the U2 aircraft carrying Francis Gary Powers is an example showing that high-altitude flight is not impervious to surface weapons.

An example of a fairly modern surface-to-air weapon is the the SA-10 Grumble. It is available for sale on the open market. Its newer versions boast a range of 56 miles and an effective intercept ceiling of 26.82 km. This height extends well into that is recognized as near-space and exceeds the altitude of some of the vehicles forecast to operate there. Thus, the lower reaches of near-space are not beyond current conventional weapons. Near-space vehicles need to be flexible, replaceable, and cost-effective, otherwise their use may be very limited in military applications.

2.4.3 Legal Constraints

Freedom of overflight is another limitation. The legal status of the near-space regime is a gray area that is not directly addressed by treaty or policy. Near-space is not a new legal regime; the question is only whether it falls under air law, where nations claim sovereignty, or space law, where overflight rights exist. Due to lack of clear legal precedent governing the near-space regime, there is considerable disagreement over whether overflight rights exist [4].

2.5 Concluding Remarks

Near-space has become an area of exceptional interest for space professionals in the last several years. Near-space vehicles can function as surrogate satellites but offer the advantage of shorter transmission distances for relaying ground-based communication and ranges shorter than those of space-borne sensors for surveillance applications. In military, the persistent surveillance from a fixed position by near-space vehicles, in contrast to periodic snapshots from the moving platforms that satellites or airplanes provide, allows continuous collection and comparison analysis over time of terrain covered by different sensors, such as infrared, electro-optical, and hyper-spectral imagery. In commercial, the comparative advantage of near-space vehicles to other means of broadband services has been assessed by many authors [1, 2, 7]. Commercial manufacturers are proposing near-space vehicles, including fixed-wing aircraft and high-altitude aircraft, to serve as surrogate satellites at a presumably reduced cost. These efforts are driven by a desire to expand commercial high-bandwidth data services. The high cost of using satellites for that purpose has motivated the development of near-space vehicles for these commercial purposes.

Near-space vehicles can thus form an additional layer of persistence between satellites and airplanes. They improve upon the long-term persistence traditionally provided by satellites while providing the fast responsiveness of airplanes. Near-space vehicles thus offer an opportunity for developing new radar remote sensing techniques. First, they can support uniquely effective and economical operations. Second, they enable new remote sensing techniques. Third, they provide a crucial corridor for prompt regional strike. The details will be discussed in the following chapters.

References

1. Zavala, A.A., Lius, J.C.R., Antonio, D.P.J.: High-Altitude Platforms for Wiress Communications. Wiley, Hoboken (2008)
2. Grace, D., Mihael, M.: Broadband Communications via High Altitude Platforms. Wiley, Hoboken (2011)
3. Wang, W.Q.: Near-space vehicles: supply a gap between satellites and airplanes for remote sensing. IEEE Aerosp. Electron. Syst. Mag. **26**, 4–9 (2011)
4. Tomme, E.B.: The paradigm shift of effects-based space: near-space as a combat space effects enabler. http://www.airpower.au.af.mi (2009). Accessed May 2010
5. Djuknic, G.M., Freidendelds, J., Okunev, Y.: Establishing wireless communications severices via high-altitude aeronautical platforms: a concept whose time has come? IEEE Commun. Mag. **35**, 128–135 (1997)
6. Tozer, T.C., Grace, D.: High-altitude platforms for wireless communications. IEE Electron. Commun. Eng. J. **13**, 127–137 (2001)
7. Grace, D., Daly, N.E., Tozer, T.C., Burrand, A.G., Pearce, D.A.J.: Providing multimedia communications from high altitude platforms. Int. J. Sat. Commun. **19**, 559–580 (2001)
8. Avagnina, D., Dovis, F., Ghiglione, A., Mulassano, P.: Wireless networks based on high-altitude platforms for the provision of integrated navigation/communication services. IEEE Commun. Mag. **40**, 119–125 (2002)
9. Jaroslav, H., David, G., Pavel, P.: Effect of antenna power roll-off on the performance of 3G cellular systems from high altitude platforms. IEEE Trans. Aerosp. Electon. Syst. **46**, 1468–1477 (2010)
10. Fidler, F., Knapek, M., Horwath, J., Leeb, W.R.: Optical communications for high-altitude platforms. IEEE J. Sel. Top. Quantum Electron. **16**, 1058–1070 (2010)
11. Anastasopoulos, M.P., Cottis, P.G.: High altitude platform networks: a feedback suppression algorithm for reliable multicast/broadcast services. IEEE Trans. Wireless Commun. **8**, 1639–1643 (2009)
12. Celcer, T., Javornik, T., Mohorcic, M., Kandus, G.: Virtual multiple input multiple output in multiple high-altitude platform constellations. IET Commun. **3**, 1704–1715 (2009)
13. Liu, Y., Grace, D., Mitchell, P.D.: Exploiting platform diversity for QoS improvement for users with different high altitude platform availability. IEEE Trans. Wireless Commun. **8**, 196–203 (2009)
14. Holis, J., Pechac, P.: Elevation dependent shadowing model for mobile communications via high altitude platforms in built-up areas. IEEE Trans. Antenna Propag. **56**, 1078–1084 (2008)
15. Likitthanasate, P., Grace, D., Mitchell, P.D.: Spectrum etiquettes for terrestrial and high-altitude platform-based cognitive radio systems. IET Commun. **2**, 846–855 (2008)
16. White, G.P., Zakharov, Y.V.: Data communications to trains from high-altitude platforms. IEEE Trans. Vehicular. Tech. **56**, 2253–2266 (2007)

17. Karapantazis, S., Pavlidou, F.: Broadband communications via high-altitude platforms: a survey. IEEE Commun. Survey Tutorial **7**, 2–31 (2005)
18. Karapantazis, S., Pavlidou, F.: The role of high-altitude platforms in beyond 3G networks. IEEE Wireless Commun. **12**, 33–41 (2005)
19. Grace, D., Thornton, J., Chen, G., White, G.P., Tozer, T.C.: Improving the system capacity of broadband services using multiple high-altitude platforms. IEEE Trans. Wireless Commun. **4**, 700–709 (2005)
20. Lee, Y., Ye, H.: Sky station statospheric telecommunications systems, a high speed low latency switched wireless network. In: Proceedings of 17th AIAA International Communication Satellite System Conference, pp. 25–32, Yokohama, Japan (1998)
21. Ilcev, S.D.: Global Mobile Satellite Communications for Maritime, Land and Aeronautical Applications. Springer, Berlin (2005)
22. Oodo, M., Tsuji, H., Miura, R., Maruyama, M., Suzuki, M., Nishi, Y., Sasamoto, H.: Experiments on IMT-2000 using unmanned solar-powered aircraft at an altitude of 20 km. IEEE Trans. Vehicular Technol. **54**, 1278–1294 (2005)
23. Wierzbanowski, T.: Unmanned aircraft systems will provide access to the statosphere. RF Des. **60**, 12–16 (2006)
24. http://www.lindstrand.co.u. Accessed Dec 2010
25. Grace, D., Thornton, J., White, G.P., Spillard, C.L., Pearce, D.A.J., Mohoreie, M., Javornik, T., Falletti, E., Delgado-Penin, J.A., Bertran, E.: The European HeliNet broadband communications application—an update on progress. In: Proceedings of 4th Japanese Stratospheric Platform Systems Workshop, pp. 90–98, Tokyo, Japan (2003)
26. Lopresti, L., Mondin, M., Orsi, S., Pent, M.: Heliplat as a GSM base station: a feasibility study. Eur. Space Agency Spec. Publ. **447**, 581–54 (1998)
27. Grace, D., Mohorcic, M., Capstick, M.H., Pallavicini, M.B., Fitch, M.: Integrating users into the wider broadband network via high altitude platforms. IEEE Trans. Wireless Commun. **12**, 98–105 (2005)
28. Yokomaku, Y.: Overview of stratospheric platform airship R&D program in Japan. In: Proceedings of 2nd Stratospheric Platform Systems Workshop, pp. 15–23, Akron, USA (2000)
29. Lee, Y.G., Kim, D.M., Yeom, C.H.: Development of Korean high altitude platform systems. Int. J. Wireless Inf. Network **13**, 31–42 (2006)
30. Jiang, B., Gao, Z.F., Shi, P., Xu, Y.F.: Adaptive fault-tolerant tracking control of near-space vehicle using Takagi-Sugeno fuzzy models. IEEE Trans. Fuzzy Syst. **18**, 1000–1007 (2010)
31. Hu, S.G., Fang, Y.W., Xiao, B.S., Wu, Y.L., Mou, D.: Near-space hypersonic vehicle longitudinal motion control based on Markov jump system theory. In: Proceedings of 8th World Congress Intelligent Control Automation, pp. 7067–7072, Jian, China (2010)
32. Ji, Y.H., Zong, Q., Dou, L.Q., Zhao, Z.S.: High-order sliding-mode observer for state estimation in a near-space hypersonic vehicle. In: Proceedings of 8th World Congress Intelligent Control Automation, pp. 2415–2418, Jian, China (2010)
33. He, N.B., Jiang, C.S., Gao, Q., Gong, C.L.: Terminal sliding mode control for near-space vehicle. In: Proceedings of 29th Chinese Control Conference, pp. 2281–2283, Beijing, China (2010)
34. Wang, W.Q., Cai, J.Y., Peng, Q.C.: Near-space microwave radar remote sensing: potential and challenge analysis. Remote Sens. **2**, 717–739 (2010)
35. Willis, N.J.: Bistatic Radar. Artech House, Norwood, MA (1995)

Chapter 3
Near-Space Vehicles in Passive Remote Sensing

Abstract There is a region in near-space where the average wind is less than 10 m/s; hence, persistent coverage and high flying speed can be obtained for vehicles operating in this region. For this reason, near-space has received much attention in the recent years. In this chapter, we consider mainly the role of near-space vehicles in passive remote sensing applications from a top-level system description.

Keywords Near-space · Passive remote sensing · Surveillance and reconnaissance · Synthetic aperture radar (SAR) · Bistatic SAR · Persistently monitoring

3.1 Near-Space Vehicles in Passive Surveillance

Spaceborne and airborne radars have played an important role in surveillance and reconnaissance; however, regardless of how well they are envisioned and employed [1], it is impossible for limited spaceborne radars to provide a persistent coverage for an area of interest, because generally speaking, most LEO satellites have a specific target in view for less than ten minutes at a time and revisit the same site infrequently. Similarly, persistent coverage is also impossible for airborne radars. By contrast, persistent coverage is possible for near-space vehicles. Near-space vehicles, especially free-floaters, are inherently survivable [2]. Moreover, several types of such vehicles are being studied, developed, or employed [3–6]. Their advantages provide a potential to passive surveillance applications. However, little work on near-space vehicles in radar and navigation applications has been reported [7–10].

3.1.1 System Configuration

The simplest and most natural application of near-space vehicles for surveillance is the passive radar system [11]. This system involves placing a passive receiver inside near-space vehicles and utilizing opportunistic illuminators such as global

W.-Q. Wang, *Near-Space Remote Sensing*,   19
SpringerBriefs in Electrical and Computer Engineering,
DOI: 10.1007/978-3-642-22188-0_3, © Wen-Qin Wang 2011

Fig. 3.1 System
configuration with
two-channel receiver

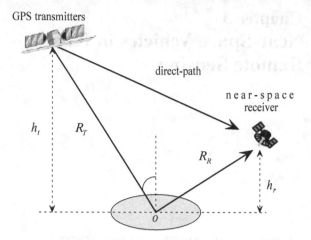

positioning system (GPS) receivers and spaceborne or airborne imaging radars. The possibility of using GPS signals reflected off the sea surface and received by air- or spaceborne sensors have been proved by several authors [12–15]. This passive remote sensing can provide persistent monitoring without significantly impacting the normal day-to-day human activities [16].

As shown in Fig. 3.1, the passive receiver contains two channels: (1) one channel is used to receive the scattered signals with which detecting of targets is attempted; and (2) the other channel is employed for receiving the direct-path signals, which are used as the reference signal for subsequent matched filtering. This system offers two other advantages [17–19]: the first advantage is the potential of bistatic observation and the second is regional persistence.

Another configuration using two or more passive receivers is also implementable. In this case, each receiver performs its own matched filtering. The total results can then be combined in some manner to provide a consistent detection or location. Additionally, a configuration using a single-channel receiver is also feasible. In this case, the received signals would contain the energy from both the direct-path channel and the scattered channel. Once they are separated, matched filtering can then be obtained successfully.

3.1.2 Signal Models

As an example, we consider the GPS L1 signal [20]

$$s_{L1}(t) = \sqrt{2P_I}d(t)c(t)\cos{(2\pi f_{L1}t + \theta_{L1})} + \sqrt{2P_Q}d(t)p(t)\sin(2\pi f_{L1}t + \theta_{L1})$$
$$(3.1)$$

where P_I and P_Q are the respective carrier power for the in-phase and quadrature-phase components, $d(t)$ is the 50 bps data modulation, $c(t)$ and $p(t)$ are the respective

Fig. 3.2 Functional blocks
of the correlation processor

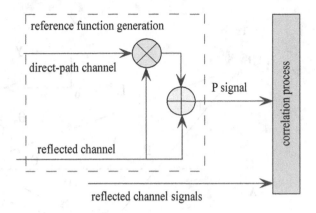

C/A and P pseudorandom code waveforms, f_{L1} is the L_1 carrier frequency in Hz
per second, and θ_{L1} is a common phase shift in radians. The quadrature carrier P_Q
is approximately 3 dB less than P_I. Then, the received GPS scattered signal is

$$x_r(t) = \alpha_r s_{L1}(t - \tau_r) \tag{3.2}$$

where α_r and τ_r are the attenuation of the scattered signal and the corresponding
delays in time, respectively.

If matched filtering is applied with a reference function of the direct-path signal,
the P codes will be masked by the C/A codes at the matched filtering output. This
problem can be resolved by filtering out the C/A components in the direct-path channel. The corresponding process is shown in Fig. 3.2. However, only a very limited
resolution can be obtained from this method. Practically, GPS provides three types of
measurements: pseudorange, carrier phase, and Doppler. Pseudorange measurement
is based on the correlation of a satellite's transmitted code and the local receiver's
reference code, which has not been corrected for errors in synchronization between
the transmitter's clock and the receiver's clock; hence, it is a time error biased measurement. Carrier phase measurement is the difference between the phases of the
receiver generated carrier signal and the carrier signal received from a satellite at the
measurement instant. Carrier phase gives more precise measurements than pseudo-
ranges, by estimating its instantaneous rate, or accumulated phase.

Taking carrier phase measurement as an example, if the phase of the received
carrier signal for the GPS satellite is denoted as ϕ_S and the phase of the reference
carrier signal generated by the receiver as ϕ_R, then there are

$$\phi_S(t_S) = \phi_S(T_S) + \phi_S(\Delta t_S), \quad \phi_R(t_R) = \phi_R(T_R) + \phi_R(\Delta t_R) \tag{3.3}$$

The time is an epoch considered from an initial epoch. The corresponding carrier
beat phase is

$$\phi_R^M(t_R) = \phi_S(t_S) - \phi_R(t_R) = \phi_S(T_S) - \phi_R(T_R) + f_{L1}\Delta t_S - f_{L1}\Delta t_R \tag{3.4}$$

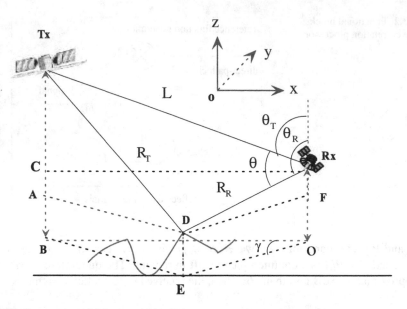

Fig. 3.3 Geometry model for target location

The term f_{L1}/c_0 with c_0 the speed of light can be used to convert the geometric distance ρ into cycles. By accounting for all error sources affecting the carrier phase measurements, the observation in cycles can be represented by

$$\phi_M = \rho + \delta\rho + c_0(\Delta t_S - \Delta t_R) - d_{ion} + \varepsilon_\phi + \lambda N \tag{3.5}$$

where $\delta\rho$ is the orbital error, d_{ion} is the ionospheric advance, ε_ϕ is the receiver phase noise, λ is the wavelength of the GPS carrier, and N is the number of integer cycles. The details can be found in many GPS textbooks. In this way, the transmitter-to-receiver distance and transmitter-target-receiver distance can be determined in a high-precision manner.

3.1.3 Target Location

To determine the instantaneous target position, we consider Fig. 3.3, which shows the passive bistatic radar geometry relations for target location. There are

$$R_{RC} = L \cos(90° - \theta_T) \tag{3.6}$$

$$R_{DF} = R_R \cos(90° - \theta_R) \tag{3.7}$$

$$\sin(\angle TDA) = \frac{h_t - h_r + R_R \sin(90° - \theta_R)}{R_T} \tag{3.8}$$

where h_t and h_r are the transmitter altitude and the receiver altitude, respectively. The other variables are illustrated in Fig. 3.3. We can get

$$\cos(\theta) = \frac{R_T^2 - L^2 - R_R^2}{2LR_R} \qquad (3.9)$$

$$\cos(\gamma) = \frac{L^2 \sin^2(\theta_T) + R_R^2 \sin^2(\theta_R) - R_T^2 \cos^2(\angle TDA)}{2LR_R \sin(\theta_T) \sin(\theta_R)}. \qquad (3.10)$$

Since there are $L \gg R_R$ (target-to-receiver distance) and R_T (transmitter-to-target distance) $\gg R_R$, we have

$$\cos(\angle TDA) \approx \frac{\gamma}{\theta} \qquad (3.11)$$

where the γ and θ are illustrated in Fig. 3.3.

Suppose

$$\Delta R = R_T + R_R - L \qquad (3.12)$$

we can get

$$\cos^{-1}\left(\frac{\Delta R(2L + \Delta R) - 2R_R(L + \Delta R)}{2LR_R}\right)$$

$$= \frac{\cos^{-1}\left(\frac{L^2 \sin^2(\theta_T) + R_R^2 \sin^2(\theta_R) - (L - \Delta R - R_R)^2\left(1 - \left(\frac{H - R_R \cos(\theta_R)}{L + \Delta R - R_R}\right)^2\right)}{2LR_R \sin(\theta_T) \sin(\theta_R)}\right)}{\left(1 - \left(\frac{H - R_R \cos(\theta_R)}{L + \Delta R - R_R}\right)^2\right)^{1/2}}$$

$$(3.13)$$

As the ΔR and L can be determined with GPS carrier phase measurement and correlation processor, the target-to-receiver distance R_R can be calculated from this equation. Next, the instantaneous target position (x_s, y_s, z_s) can be estimated by

$$(x_s, y_s, z_s) = (x_r - R_R \cos(90° - \theta_R) \cos(\varepsilon),$$
$$y_r - R_R \cos(90° - \theta_R) \sin(\varepsilon), z_r - R_R \sin(90° - \theta_R)) \qquad (3.14)$$

where (x_r, y_r, z_r) is the location of the receiver.

3.1.4 Power Budget Analysis

Using GPS satellite as an illuminator for near-space vehicle-borne passive bistatic radar presents a problem of signal detectability, since the reflected GPS signal is very weak. After coherent bistatic data processing, the final signal-to-noise ratio (SNR) can be represented by [21]

Fig. 3.4 Example NESZ
performance

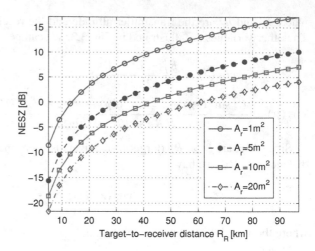

Target–to–receiver distance R_R [km]

$$SNR_{\text{radar}} = \Pi_0 \cdot \frac{A_r \sigma_0}{4\pi R_R^2} \cdot \frac{1}{K T_0 F_n} \cdot \frac{R_R T_s}{\lambda R_T} \cdot \eta \tag{3.15}$$

where $\Pi_0 \approx 3 \times 10^{-14}\,\text{Wt/m}^2$, σ_0 is the radar cross-section (RCS) parameter, K is the Boltzmann constant, T_0 is the system noise temperature, F_n is the noise figure, A_r is the effective receiver antenna area, and η is the loss factor.

A quantity directly related to radar detection performance is the noise equivalent sigma zero (NESZ), which is the mean RCS necessary to produce an SNR_{radar} of unity. The NESZ can be interpreted as the smallest target RCS, which is detectable by the radar system against thermal noise. Setting $SNR_{\text{radar}} = 1$, Eq. 3.15 gives

$$NESZ = \frac{4\pi R_r^2 F_n K T_0}{\Pi_0 A_r T_s \eta} \tag{3.16}$$

As an example, assuming a typical system with the following parameters: $T_s = 1{,}000\,\text{s}$, $T_0 = 300$, $\eta = 0.5$, and $F_n = 2\,\text{dB}$, the calculated NESZ is illustrated in Fig. 3.4. This result shows that an RCS requirement comparable to the current radar systems is possible for near-space vehicle-borne passive bistatic radar systems.

3.2 Near-Space Vehicles in Passive BiSAR Imaging

The use of the GPS illuminator has the advantages of entire planet coverage and simple transmitter-receiver synchronization [22], although high-resolution imaging cannot be obtained due to its limited signal bandwidth [23, 24]. In this section, we further discuss near-space vehicles in passive bistatic synthetic aperture radar (BiSAR) imaging when spaceborne or airborne radar is used as the opportunistic transmitter. Although the passive near-space vehicle-borne receiver may be stationary, an aperture synthesis can still be obtained by the transmitter's motion.

Fig. 3.5 General geometry
of near-space vehicle-borne
passive BiSAR

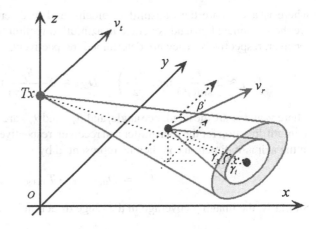

BiSAR has received much recognition in the recent years [25]. But the published BiSAR systems are mainly azimuth-invariant configurations [26], for example the "Tandem" configuration where the transmitter and receiver move one after another with the same trajectory and the "translational invariant" configuration where the transmitter and receiver move along parallel trajectories with the same velocity. As near-space vehicle-borne receivers and spaceborne or airborne transmitters may have unequal velocities and/or unparallel trajectories, here, we consider mainly the azimuth-variant processing challenges and potential solutions.

3.2.1 System Imaging Performance

We consider one general near-space vehicle-borne passive BiSAR imaging geometry, as shown in Fig. 3.5. The transmitter and passive receiver fly on a straight but nonparallel path with a constant but unequal velocity. The azimuth time is chosen to be zero at the composite beam center crossing time of the reference target.

3.2.1.1 Imaging Time and Imaging Coverage

For BiSAR imaging, both the transmit and receive antennas are steered to obtain an overlapping beam on the ground, and the exposure of a target is governed by this composite beam pattern. Thus, it is necessary to analyze the imaging time and imaging coverage. A space-time diagram-based approach was investigated in [27], but a rectangular approximation is employed. In fact, the whole imaging time of the near-space vehicle-borne BiSAR is determined by

$$T_{\text{image}} = \frac{D_{a,t} + D_{a,r}}{|v_t - v_r|} \tag{3.17}$$

where v_t and v_r are the transmitter velocity and receiver velocity, $D_{a,t}$ and $D_{a,r}$ are the illuminated ground coverage in azimuth direction for the transmitter and the receiver, respectively. They are determined, respectively, by

$$D_{a,t} \approx 2\frac{h_t}{\sin(\gamma_t)}\tan\left(\frac{\theta_{a,t}}{2}\right), \quad D_{a,r} \approx 2\frac{h_r}{\sin(\gamma_r)}\tan\left(\frac{\theta_{a,r}}{2}\right) \tag{3.18}$$

where γ_t and γ_r are the incidence angle, and $\theta_{a,t}$ and $\theta_{a,r}$ are the antenna beamwidth in azimuth direction for the transmitter and receiver, respectively. The imaging coverage in the azimuth direction can then be represented by

$$L_{az} = D_{a,r} - v_r \cdot T_{\text{image}} \tag{3.19}$$

Similarly, the imaging coverage in the range direction is determined by

$$L_{ra} = D_{r,r} \approx 2\frac{h_r}{\sin(\gamma_r)}\tan\left(\frac{\theta_{r,r}}{2}\right) \tag{3.20}$$

where $D_{r,r}$ and $\theta_{r,r}$ are the imaging ground coverage and antenna beamwidth in range direction for the receiver, respectively.

3.2.1.2 Range Resolution

The instantaneous range history of the transmitter and receiver to an arbitrary point target $(x, y, 0)$ is

$$R = \sqrt{(x - x_t)^2 + (y - y_t)^2 + h_t^2} + \sqrt{(x - x_r)^2 + (y - y_r)^2 + h_r^2} \tag{3.21}$$

where (x_t, y_t, h_t) and (x_r, y_r, h_r) are the coordinates of the transmitter and the receiver, respectively. We then have

$$\nabla R = \frac{\partial R}{\partial x}\mathbf{i}_x + \frac{\partial R}{\partial y}\mathbf{i}_y$$
$$= [\sin(\alpha_t)\cos(\zeta_t) + \sin(\alpha_r)\cos(\zeta_r)]\mathbf{i}_x + [\sin(\zeta_t) + \sin(\zeta_r)]\mathbf{i}_y \tag{3.22}$$

where $\alpha_t = \alpha_t(x)$ and $\alpha_r = \alpha_r(x)$ are the instantaneous looking-down angles, $\zeta_t = \zeta_t(x, y; y_t)$ and $\zeta_r = \zeta_r(x, y; y_r)$ (y_t, y_r is the instantaneous location in y-direction) are the instantaneous squint angles. There is

$$|\nabla R| = \sqrt{[\sin(\alpha_t)\cos(\zeta_t) + \sin(\alpha_r)\cos(\zeta_r)]^2 + [\sin(\zeta_t) + \sin(\zeta_r)]^2} \tag{3.23}$$

As the range resolution of a monostatic SAR is $c_0/2B_r$ with B_r the transmitted signal bandwidth, the range resolution (in x direction) can then be derived as

$$\rho_r = \frac{c_0/B_r}{|\nabla R|} \cdot \frac{1}{\sin(\zeta_{xy})} \tag{3.24}$$

with

$$\zeta_{xy} = \arctan\left(\frac{\sin(\zeta_t) + \sin(\zeta_r)}{\sin(\alpha_t)\cos(\zeta_t) + \sin(\alpha_r)\cos(\zeta_r)}\right) \qquad (3.25)$$

Thus, we have

$$\rho_r = \frac{c_0/B_r}{[\sin(\alpha_t)\cos(\zeta_t) + \sin(\alpha_r)\cos(\zeta_r)]} \qquad (3.26)$$

We can conclude that the range resolution is determined by not only the transmitted signal bandwidth, but also the specific BiSAR configuration geometry.

3.2.1.3 Azimuth Resolution

At an azimuth time τ, the range sum to an arbitrary reference point is

$$R(\tau) = \sqrt{R_{t0}^2 + (v_t\tau)^2} + \sqrt{R_{r0}^2 + (v_r\tau)^2} \qquad (3.27)$$

where R_{t0} and R_{r0} are the closest ranges to a given point target when the transmitter and receiver move along their trajectories, respectively. The Doppler chirp rate is derived as

$$k_d(\tau) = -\frac{1}{\lambda} \cdot \frac{\partial^2 R(\tau)}{\partial^2 \tau} \approx -\frac{1}{\lambda} \cdot \left[\frac{v_t^2}{R_{t0}}\cos\left(\frac{v_t\tau}{R_{t0}}\right) + \frac{v_r^2}{R_{r0}}\cos\left(\frac{v_r\tau}{R_{r0}}\right)\right] \qquad (3.28)$$

The synthetic aperture time is determined by [28]

$$T_s = \min\left\{\frac{\lambda R_{t0}}{L_t v_t}, \frac{\lambda R_{r0}}{L_r v_r}\right\} \qquad (3.29)$$

where L_t and L_r are the transmit antenna length and receive antenna length, respectively.

If $\frac{\lambda R_{t0}}{L_t v_t} > \frac{\lambda R_{r0}}{L_r v_r}$, the corresponding Doppler bandwidth is

$$B_d(\tau) = \int_{-T_s/2}^{T_s/2} |k_d(\tau)| d\tau = 2\left[\frac{v_t}{\lambda}\sin\left(\frac{\lambda R_{r0}v_t}{2D_r R_{t0}v_r}\right) + \frac{v_r}{\lambda}\sin\left(\frac{\lambda}{2D_r}\right)\right] \qquad (3.30)$$

The azimuth resolution can then be expressed as

$$\rho_a = \frac{v_{eq}}{B_d(\tau)} = \frac{v_{eq}}{2\left[\frac{v_t}{\lambda}\sin\left(\frac{\lambda R_{r0}v_t}{2L_r R_{t0}v_r}\right) + \frac{v_r}{\lambda}\sin\left(\frac{\lambda}{2L_r}\right)\right]} \qquad (3.31)$$

Table 3.1 Example BiSAR configuration parameters using near-space vehicle-borne receiver

Parameters	Transmitter A	Receiver A	Transmitter B	Receiver B
Carry frequency (GHz)	5.33	5.33	1.25	1.25
Flying altitude (km)	800	20	10	20
Flying velocity (m/s)	7,450	5	100	5
Signal bandwidth (MHz)	16	16	500	500
Beam incidence angle (°)	30	60	60	60
Beamwidth in range (°)	2.3	15	15	15
Beamwidth in azimuth (°)	0.28	15	13.75	13.75
Imaging time (s)	1.8670		85.3248	
Imaging coverage (x,y)(km)	(6.0808, 6.0714)		(2.3448, 3.0404)	

where

$$v_{eq} = \sqrt{v_t^2 + v_r^2 - 2v_t v_r \cos(\pi - \beta')} \tag{3.32}$$

is the equivalent velocity, β' is defined as the angle between the transmitter velocity vector and the receiver velocity vector. Similarly, if $\frac{\lambda R_{t0}}{L_t v_t} < \frac{\lambda R_{r0}}{L_r v_r}$, there is

$$\rho_a = \frac{v_{eq}}{2\left[\frac{v_r}{\lambda}\sin\left(\frac{\lambda R_{t0}v_r}{2L_t R_{r0}v_t}\right) + \frac{v_t}{\lambda}\sin\left(\frac{\lambda}{2L_t}\right)\right]} \tag{3.33}$$

From Eqs. 3.31 and 3.33 we can see that, unlike the azimuth resolution of general monostatic SAR determined only by the azimuth antenna length, here the azimuth resolution will be impacted by its BiSAR configuration geometry and velocity. When $R_{t0} = R_{r0}$, $v_t = v_r$ and $L_t = L_r$, there is $\rho_a = L_t/2 = L_r/2$. In this case, it is just a monostatic SAR. If $v_t = 0$ or $v_r = 0$ and $L_t = L_r$, the azimuth resolution is found to be $\rho_a = L_t = L_r$. This case is just a fixed-transmitter or fixed-receiver BiSAR [29]. Note that there is a local minimum for the azimuth resolution, which is $L_r/2$ or $L_t/2$ (depending on whether $\frac{\lambda R_{t0}}{L_t v_t} > \frac{\lambda R_{r0}}{L_r v_r}$ or $\frac{\lambda R_{t0}}{L_t v_t} < \frac{\lambda R_{r0}}{L_r v_r}$).

3.2.1.4 Simulation Results

To obtain quantitative evaluation results, we consider several typical configurations of near-space vehicle-borne BiSAR systems. Table 3.1 gives the simulation para-meters with a near-space vehicle-borne receiver. The spaceborne Envisat SAR [30] and airborne radar are assumed as the transmitters in Case A (for transmitter A and receiver B) and Case B (for transmitter B and receiver B), respectively. We note that an imaging coverage size of dozens of square kilometers (the size is from 2×3 km to 6×6 km in the simulation examples) can be obtained. Figure 3.6 gives the example range resolution results. It is noted that the range resolution has geometry-variant characteristics, which depend on not only the slant range

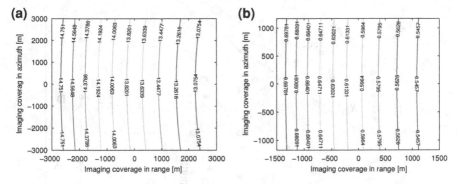

Fig. 3.6 Range resolution of the example near-space vehicle-borne BiSAR configurations. **a** *Case A* Transmitter A and Receiver A. **b** *Case B* Transmitter B and Receiver B

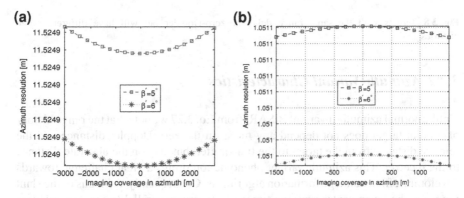

Fig. 3.7 Azimuth resolution of the example near-space vehicle-borne BiSAR configurations. **a** *Case A* Transmitter A and Receiver A. **b** *Case B* Transmitter B and Receiver B

but also the azimuth range. The range resolution degrades with the increase of azimuth range displaced from the scene center. To ensure a consistent range resolution, the imaged scene coverage should be limited or a long slant range should be designed.

Figure 3.7 gives the example azimuth resolution results. It is noted that the angle (β') between the transmitter velocity vector and receiver velocity vector has also an impact on the azimuth resolution. Additionally, we note that the azimuth resolution between scene center and scene edge has a small performance difference. This phenomenon is caused by the change in azimuth time. Further simulations show that, if the synthetic aperture time is determined primarily by the receiver, and suppose the transmitter's velocity is constant, the azimuth resolution will degrade with the increase in the receiver's velocity. However, if the synthetic aperture time is determined primarily by the transmitter, and suppose the transmitter's velocity is constant, the azimuth resolution will be improved with the increase in the near-space vehicle-borne receiver's velocity.

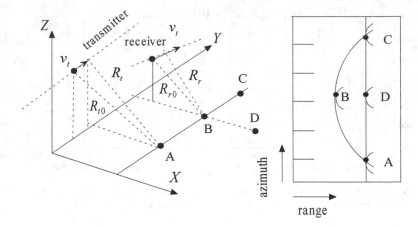

Fig. 3.8 Targets with the same range delay at zero-Doppler but will have different range histories

3.2.2 Azimuth-Variant Characteristics

Unlike normal azimuth-invariant BiSAR, from Eq. 3.27 we note that the range history of a point target does not depend any more on the zero-Doppler distance and the relative distance from the target to the transmitter, but also on the absolute distance to the receiver. This azimuth-variant phenomenon brings a great challenge towards developing efficient image formation algorithms. Consequently, the locus of the slant ranges at the beam center crossing times of all targets parallel to the azimuth axis follow an approximate hyperbola.

We consider a typical geometry shown in Fig. 3.8 (*left*), where the (x, y) plane is locally tangent to the surface of the earth. The targets are assumed to lie on this plane, and the transmitter velocity vector is parallel to the y-axis. Figure 3.8 (*right*) shows the trajectories of three targets A, B, and C, as well as the hyperbolic locus. Because of the hyperbolic locus, after linear range cell mitigation correction the targets such as A, C and D are at the same range gate. This is different from the monostatic case, which has a linear locus instead of hyperbolic.

Consequently, the general range Doppler image formation algorithm cannot handle this problem in a high-precision manner. Moreover, there is a range ambiguity that does not exist in monostatic cases. It can be observed from Fig. 3.8 that two or more targets (e.g., A, C and D) located at different positions can have the same range delay at zero-Doppler but will have different range histories (curvature). Similar phenomena have been investigated in [31], where a ground-based stationary receiver is assumed. For the near-space vehicle-borne BiSAR this problem becomes being more complex, since the spacebore or airborne transmitter follows a rectilinear trajectory, while the near-space vehicle-borne receiver follows also a rectilinear trajectory but with a different velocity.

The situation becomes more complicated for unflat digital-earth model (DEM) topography. In monostatic or azimuth-invariant BiSAR systems, scene topography can be ignored in developing image formation algorithms because the measured range delay is related to the double target distance and the observed range curvature; hence, topography is only used to project the compressed image which is in slant range to the ground range. However, for the near-space vehicle-borne BiSAR it is mandatory to know both the transmitter-to-target distance and target-to-receiver distance to properly focus on its raw data, which clearly depend on the target height. In this case, the conventional imaging algorithms such as Chirp-Scaling and wavenumber-domain, may not be suitable to accurately focus on the collected data.

3.2.3 Two-Dimensional Spectrum Model

The two-dimensional spectrum model is useful for developing an effective image formation algorithm. We consider the demodulated near-space BiSAR signal

$$S(f_r, f_a) = P(f_r) \cdot \int\limits_{-\infty}^{\infty} \exp\left(-j2\pi \frac{f_r + f_0}{c_0} R(\tau) - j2\pi f_a \tau\right) d\tau \qquad (3.34)$$

where f_r, f_a and f_0 denote the range-frequency, azimuth-frequency, and carrier frequency, respectively. Loffeld et al. [32] divided the bistatic phase history into two quasi-monostatic phase histories and expended them into Taylor series around the individual points of stationary phase

$$\Phi_t(\tau) = -2\pi \left[(f_r + f_0)\frac{R_t(\tau)}{c_0} - f_a \tau\right]$$

$$\approx \Phi_t(\tau_t^*) + \Phi_t'(\tau_t^*)(\tau - \tau_t^*) + \frac{1}{2}\Phi_t''(\tau_t^*)(\tau - \tau_t^*)^2 + \cdots \qquad (3.35)$$

$$\Phi_r(\tau) = -2\pi \left[(f_r + f_0)\frac{R_r(\tau)}{c_0} - f_a \tau\right]$$

$$\approx \Phi_r(\tau_r^*) + \Phi_r'(\tau_r^*)(\tau - \tau_r^*) + \frac{1}{2}\Phi_r''(\tau_r^*)(\tau - \tau_r^*)^2 + \cdots \qquad (3.36)$$

Let

$$\Phi_t'(\tau_t^*) = 0, \quad \Phi_r'(\tau_r^*)(\tau - \tau_r^*) = 0 \qquad (3.37)$$

There is

$$\tau^* = \frac{\Phi_t''(\tau_t^*)\tau_t^* + \Phi_r''(\tau_r^*)\tau_r^*}{\Phi_t''(\tau_t^*) + \Phi_r''(\tau_r^*)} \qquad (3.38)$$

(a)
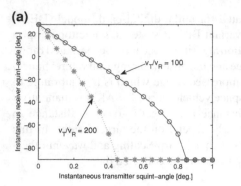

(b)
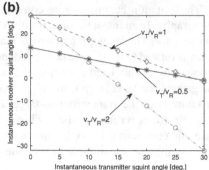

Fig. 3.9 The constraints of the Loffeld's BiSAR two-dimensional spectrum model. **a** In small squint-angle cases. **b** In small difference between v_T and v_R

In this way, one two-dimensional BiSAR spectrum model is constructed in [32]. Unfortunately, Eq. 3.38 is constrained by

$$|\tau_t^* - \tau_{t0}|^2 \ll \frac{2R_{t0}^2}{7v_t^2}, \quad |\tau_r^* - \tau_{r0}|^2 \ll \frac{2R_{r0}^2}{7v_r^2} \tag{3.39}$$

where τ_{t0} and τ_{r0} denote the shortest transmitter-to-target distance and receiver-to-target distance, respectively. Figure 3.9 shows that this model can only be used in small squint angle or small difference between the transmitter velocity and the receiver velocity.

Another model using series expansion to express azimuth time as a function of azimuth frequency during azimuth Fourier transform was proposed in [33], and a chirp scaling algorithm was developed based on this model [34]. The accuracy is controlled by keeping enough terms in the power series, but only the first several series can be used. Using the Fresnel approximation, one 2-D BiSAR spectrum model was derived in [35]. This model can be used only in small squint angle. Moreover, it has no advantages in approximate errors. In summary, most BiSAR spectrum models have to use some approximations by keeping enough terms in the power series, but cannot be used in the BiSAR with large difference between transmitter-to-target distance and target-to-receiver distance or large squint angle, like the near-space vehicle-borne azimuth-variant BiSAR.

To develop an effective 2-D spectrum model for the near-space vehicle-borne BiSAR, we consider the azimuth-variant BiSAR geometry shown in Fig. 3.10 and the range history sum to an arbitrary point given in Eq. 3.27. Taking spaceborne transmitter as an example and supposing the synthetic aperture time is T_s, since there is

$$v_t \gg v_r, \quad R_{t0} \gg v_r\tau, \quad \tau \in [-T_{s/2}, T_{s/2}] \tag{3.40}$$

we then have

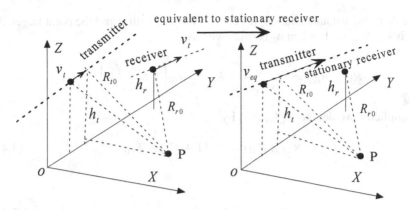

Fig. 3.10 Azimuth-variant BiSAR geometry and its equivalent model

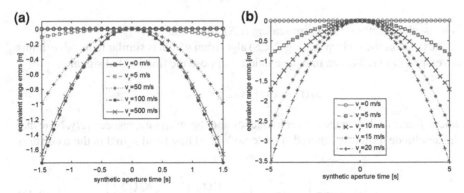

Fig. 3.11 Range errors caused by the equivalent BiSAR geometry model. **a** $R_{t0} = 800\,\text{km}$, $v_t = 7600\,\text{m/s}$. **b** $R_{t0} = 15\,\text{km}$, $v_t = 100\,\text{m/s}$

$$R(\tau) = \sqrt{R_{t0}^2 + (v_t\tau)^2} + \sqrt{R_{r0}^2 + (v_r\tau)^2} \approx \sqrt{R_{t0}^2 + (v_{\text{eq}}\tau)^2} + R_{r0} \qquad (3.41)$$

where $v_{eq} = v_t + v_r$ is the equivalent velocity between the transmitter and the receiver. Figure 3.11 gives the range errors caused by the equivalent BiSAR model. The phase errors that are smaller than $\pi/4$ can be ignored for subsequent image formation processing algorithms. We thus conclude that this equivalent BiSAR model is feasible for near-space vehicle-borne BiSAR systems.

3.2.4 Image Formation Processing

The equivalent instantaneous azimuth Doppler frequency f_a is represented by

$$f_a = \frac{v_{\text{eq}}}{\lambda} \sin(\phi_t) \qquad (3.42)$$

where ϕ_t is the instantaneous angle between the transmitter and the point target. As there is $\lambda f_a / v_{eq} \ll 1$, when ϕ_t is small we have

$$R(f_a; R_{t0}) = \frac{R_{t0}}{\cos(\phi_t)} + R_{r0} \approx R_{t0}\left[1 + \frac{\lambda^2 f_a^2}{2v_{eq}^2}\right] + R_{r0} \qquad (3.43)$$

For simplicity, we denote $R(f_a; R_{t0})$ by

$$R(f_a; R_{t0}) = R_{t0}[1 + C_s] + R_{r0} \qquad (3.44)$$

where

$$C_s = \frac{1}{\sqrt{1 - (\lambda f_a / v_{eq})^2}} - 1 \qquad (3.45)$$

is the scaling factor of the chirp scaling (CS) algorithm.

Next, a nonlinear chirp scaling (NCS) algorithm which is similar to the algorithms developed in [31, 34] can then be applied. Suppose the transmitted signal is

$$s_t(t) = \exp[j\pi(2f_c t + k_r t^2)] \qquad (3.46)$$

where f_c and k_r denote the carrier frequency and the chirp rate, respectively. Note that the amplitude terms are ignored. The demodulated baseband signal in the receiver is

$$s_r(t) = \exp\left\{-j\pi\left[k_r\left(t - \frac{R(\tau)}{c_0}\right)^2 + 2\frac{R_b(\tau)}{c_0}\right]\right\}. \qquad (3.47)$$

Applying one Fourier transform (it is often implemented with fast Fourier transform (FFT)) to azimuth time τ yields

$$S_r(t, f_a) = \exp\left\{-j\pi k_{eq}\left[t - \frac{R(f_a; R_{t0})}{c_0}\right]^2\right\}$$
$$\times \exp\left(-j\pi \frac{\lambda R_{b0} f_a^2}{v_{eq}}\right) \cdot \exp\left(-j2\pi f_a \frac{y_p}{v_{eq}}\right) \qquad (3.48)$$

with

$$k_{eq} = \frac{1}{\frac{1}{k_r} - \frac{\lambda R_{t0}}{c_0^2} \cdot \frac{(\lambda f_a / v_{eq})^2}{\left[\sqrt{1 - (\lambda f_a / v_{eq})^2}\right]^3}} \qquad (3.49)$$

where $(x_p, y_p, 0)$ is the point target's coordinate, and R_{b0} is the smallest bistatic range.

Chirp scaling processing with the phase term

$$\Phi_1 = \exp\left\{-j\pi k_{eq}\left(\frac{1}{\sqrt{(1-(\lambda f_a/v_{eq})^2)}}-1\right)\left[t-\frac{R(f_a;R_{bref})}{c_0}\right]\right\} \qquad (3.50)$$

where R_{bref} is the reference range. Next, they are transformed into two-dimensional frequency-domain through range FFT. After range compression with the phase term

$$\Phi_2 = \exp\left\{j\pi\frac{k_{eq}}{1+C_s}f_r^2\right\}\cdot\exp\left(j\pi f_r\frac{\lambda^2 R_{bref}f_a^2}{c_0 v_{eq}^2}\right) \qquad (3.51)$$

a range inverse FFT (IFFT) is applied. To compensate the effects of azimuth-variant Doppler chirp rate, phase filtering is further applied before azimuth IFFT [31]

$$\Phi_3 = \exp\left\{j\frac{\pi\lambda^3 R_{b0}^4 f_a^4}{3L_{az}v_{eq}^6 T_{image}^4}\right\} \qquad (3.52)$$

Finally, the Doppler chirp rate is corrected by

$$\Phi_4 = \exp\left\{j\pi\left[\frac{2v_{eq}^2}{\lambda L_{az}}-\frac{v_{eq}^2}{\lambda R_{b0}}\right]\tau^2 + j\pi\left[\frac{\left(\frac{2v_{eq}^2}{\lambda L_{az}}-\frac{v_{eq}^2}{\lambda R_{b0}}\right)2R_{b0}}{T_{image}^2 L_{az}}\right]\tau^4\right\} \qquad (3.53)$$

Applying one Fourier transform in the azimuth direction and multiplying a phase compensation term

$$\Phi_5 = \exp\left[-j\pi\frac{2v_{eq}^2}{\lambda L_{az}}f_a^2\right]\cdot\exp\left[\frac{4\pi k_{eq}C_s(1+C_s)(R_{b0}-R_{bref})^2}{c_0}\right]$$

$$\times\exp\left\{-j\pi\left[\frac{1}{\left(\frac{2v_{eq}^2}{\lambda L_{az}}\right)^4 + \left(\frac{v_{eq}^2}{\lambda R_{b0}}\right)^4}-\frac{1}{48v_{eq}^6 T_{image}^2}\right]f_a^4\right\} \qquad (3.54)$$

The focused BiSAR image can then be obtained by an azimuth inverse Fourier transform. The whole processing step is illustrated in Fig. 3.12.

To evaluate the performance of the derived imaging algorithm, the two example BiSAR data from five point targets are simulated using the parameters listed in Table 3.1. A distance separation of five times the range/azimuth resolution is assumed in the simulation examples. Additionally, $\zeta_{t0}(\tau=0)=0$ and $\zeta_{r0}(\tau=0)=0$ are supposed. After processed by the equivalent velocity model and NCS combined image formation algorithm, the simulation results are given in Fig. 3.13. Note that, as the reference range used in the imaging algorithm is chosen to be the central target range, the focusing performance of the central target is better than that of the edge target. Even so, from the results we can conclude that the five point targets are well focused and the imaging performance is acceptable.

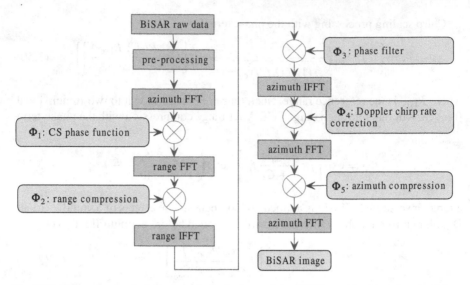

Fig. 3.12 Processing steps of the equivalent velocity model and NCS combined imaging algorithm

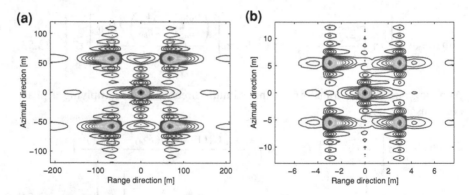

Fig. 3.13 Processing results with the image formation algorithm. **a** Case A. **b** Case B

3.3 Near-Space Vehicles in Passive Environment Monitoring

The last but not the least use of near-space vehicles in passive remote sensing is regional atmospheric environment monitoring. An example application is the near-space vehicle-borne GPS receivers for radio occultation measurements. Radio occultation measurements using GPS and a LEO spaceborne receiver can offer accurate profiles of atmospheric refractivity, pressure, water vapor, and temperature with a high vertical resolution [36, 37]. While GPS occultation data collected from space have the advantage of being of global coverage (one LEO spaceborne receiver provides about 500 globally distributed occultations per day), the occultation sampling in any particular region is relatively sparse. In contrast, near-space vehicle-borne

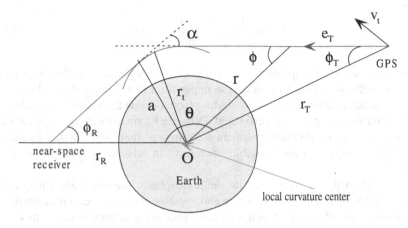

Fig. 3.14 Geometry of the near-space vehicle-borne GPS occultation

GPS receivers can offer a dense radio occultation measurement over a specific area of interest for purposes of regional weather and climate observation or coupled ocean/atmospheric process research.

As near-space vehicle-borne receivers operate in the atmosphere, the receivers can track GPS satellites as they set and rise behind the Earth's limb, therefore collecting GPS signals at both negative and positive elevations relative to the receiver's local horizon. The viewing geometry of near-space vehicle-borne GPS occultation can then be thought of as a hybrid between spaceborne and ground-based occultation geometries; hence, it combines the high vertical profiling capability of spaceborne occultation data (at least for height below the receiver) with the benefit of routinely obtaining a relatively large number of daily profiles in the regions of interest. Each occultation yields a profile of refractivity below the height of the receiver with a diffraction-limited vertical resolution. A single near-space vehicle-borne receiver with a full 360° field of view can observe several hundreds of occultations per day scattered within a radius of several hundreds of kilometers from the receiver. If multiple near-space vehicle-borne receivers can be formatted with an optimized flight plan [38], more daily occultations can be obtained over the region. This information, when combined with columnar water vapor distribution derived from upward looking ground-based receivers, and possibly moisture information from microwave remote sensing [39], is extremely useful for regional weather monitoring and hydrological and boundary layer research.

As shown in Fig. 3.14, from the knowledge of the positions of the GPS satellite and near-space vehicle-borne receiver and their clocks (obtained from other GPS measurement collected simultaneously) the delay due to the intervening media can be isolated. For spaceborne GPS occultation, when the bending is determined from the Bouguer's formula, the index of refraction can be derived from

$$n(a) = \exp\left[\frac{1}{\pi} \int_a^\infty \frac{\alpha(a')}{\sqrt{a'^2 - a^2}} \, da'\right] \qquad (3.55)$$

where a denotes the asymptote miss distance or impact parameter. The other parameters are defined as Fig. 3.14. Since the upper limit of the integral is infinity, it is necessary to have measurements of α starting from outside the atmosphere (where α vanishes) for this integral to be performed. Differently, since the near-space vehicle-borne receiver lies inside the atmosphere at radius r_R, then only the measurements of $\alpha(a')$ for $a < r_R n(r_R)$ are available; therefore, in this case Eq. 3.55 should be modified.

In fact, when the receiver is inside the atmosphere, the $\alpha(a')$ has a maximum exactly at the point that separates positive and negative elevation measurements (however, spaceborne GPS occultation has no local maximum or minimum). In this case, the maximum value for a is $a_{max} = n(r_R)r_R$. Then, from the Abel inversion scheme [40] we have

$$n(a) = \exp\left[\frac{1}{\pi} \int_a^{a_{max}} \frac{\alpha(a')}{\sqrt{a'^2 - a^2}} \, da'\right] \qquad (3.56)$$

In this way, the numerical refractions n can be obtained.

GPS occultation using near-space vehicle-borne receiver offers a simple method of obtaining vertical information about the lower atmosphere up to the receiver altitude. About several hundreds of measurements per day per receiver could be obtained in a given region, in contrast to radiosondes that are launched once or twice daily. Since vertical information is extremely important in characterizing the stability of the atmosphere and is expensive to acquire, the profiles obtainable from near-space vehicle-borne GPS receivers may provide quite useful applications such as regional weather forecasting, hydrology, surface-air exchange, and related topics on atmospheric research.

However, as a new GPS occultation technique, much further work is needed. For example, when the near-space vehicle-bore receiver is moving inside the atmosphere, separating the positive and negative elevation data will be more difficult, depending on the motion dynamics. Additionally, to assess the merits of near-space vehicle-borne GPS occultation, further experiments should be performed to investigate the possible strategies of complementing bending measurements with refractivity and to understand their impact on inversion performance.

3.4 Potential and Challenges

Near-space passive remote sensing provides many promising potential applications due to the superiorities of persistent regional coverage, robust survivability, bistatic observation [41], and low cost. Certainly, there also are several technical challenges in synchronization compensation and motion compensation.

3.4.1 Potential Applications: Homeland Security

One potential application is homeland security. To protect civilian population, mass transit, civil aviation, and critical infrastructure from terrorist attacks, significant improvement in our capability to safeguard homeland security is required, so that we can efficiently neutralize these threats without impacting the normal day-to-day human and commerce activities. To reach this aim, radar has long been used in a variety of military and civilian applications and has been an essential component of the current defense systems. Presently, many countries have their own civil aviation radar networks. These radar networks are specifically designed to ensure early warning against potential hostile targets. However, the range and nature of potential threat targets are becoming more and more diverse. The tragic event of 9/11 is but one example. Others might include missile attacks and UAVs. This means that the source of attack can emanate in a much wider variety of new and different forms, and cannot be efficiently dealt with by the current radar systems. It appears that passive near-space vehicle-borne radars using opportunistic illuminators can provide a potential solution to these problems.

Rather than emitting signals, this passive remote sensing system relies on opportunistic transmitters to detect potential targets. This allows the receiver to operate without emitting energy. This is particularly attractive for homeland security applications, because it is desirable for such a sensor to serve also other purposes such as traffic monitoring and weather prediction without influencing normal human activities.

3.4.2 Potential Applications: Persistently Disaster Monitoring

Another potential application is persistent disaster monitoring [42, 43]. The frequency of natural disasters has shown a rapid increase in the recent years [44]. Examples of this trend are related to floods, earthquakes, tsunamis, hurricanes, and forest fires [45]. The tsunami that killed thousands of people in the coastal areas of India, Indonesia, Thailand, and Srilanka has brought the awareness that we cannot regard natural disasters any more as inevitable or unavoidable and remain helpless observers. Methods and strategies along with effective techniques have to be developed to predict and tackle natural disasters. Toward this aim, immediately after that tsunami, natural disaster monitoring has received much recognition [46], but there is still a lack of feasible and practicable solutions.

Taking tsunami detection as an example, tsunamis can be triggered by earthquakes, submarine landslides, volcanic eruptions, meteorite impacts, or by a combination of all these factors. Tsunami waves have sufficient energy to cross entire oceans, traveling outward rapidly with small detectable heights in the deep ocean. But, since a tsunami wave's shoaling is near the shore, a tsunami that is visibly unnoticeable in the deep ocean will become more detectable if closer to the shore [47]. It is reported that the Sumatra tsunami had a recorded wave height of 60–80 cm in the deep ocean, but had a maximum wave height of 15 m near Banda Aceh [48]. As tsunami is a surface gravity wave with a wavelength much larger than the ocean depth, the wave develops in three stages [49]: the first stage is the formation due to the initial cause and propagation near the source, the second is the free propagation of the wave in the open ocean at large depths, and the third is the propagation of the wave in the region of continental shelf and shallow coastal waters. The wave behavior in each specific region should be understood in order to properly detect, predict, and model the tsunami wave's activity.

It is has been proved that, there is a link between the tsunami wave amplitude and the microwave RCS [44], and significant variations (a few dB) of the RCS synchronous with the sea level anomaly can be found both at C and Ku band in the geophysical data record of the altimetry satellite Jason-1. From the microwave remote sensing viewpoint, the oceanographic detectable features of a tsunami wave are summarized in [9]: (1) tsunami wave height, (2) tsunami orbital velocities, (3) tsunami-induced RCS modulations. The first two parameters inherently belong to an ocean wave, and the third is a geophysical feature that arises from complex hydrodynamic processes. The behavior of tsunami-induced RCS modulations depends upon bathymetry, meteorological factors, and sea-state. The capability of microwave radar systems to detect internal waves has been proved [50–52]. Along with tides, tsunamis are shallow water waves and have the potential to trigger internal waves. Tsunami-induced RCS modulations can be used to predict future tsunamis. This involves not only detection but also an estimate of the tsunami magnitude. Thus, near-space passive remote sensing has the possibility to detect a tsunami by comparing the pre and post quake patterns.

3.4.3 Challenges: Synchronization Compensation

In monostatic radar remote sensing, the colocated transmitter and receiver use the same stable local oscillator (STALO), the phase can decorrelate only over a very short time (about 1×10^{-3} sec). By contrast, for the near-space passive remote sensing, there is no phase noise cancellation because the transmitter and receiver use separate local oscillators. This superimposed phase noise will corrupt the received signal over the whole coherent processing time.

As only local oscillator phase noise is of interest, the passive radar can be simplified into an azimuth-only system [53]. Suppose the transmitted signal is sinusoid with phase argument

$$\Phi_T(t) = 2\pi f_T t + M\varphi_T(t). \qquad (3.57)$$

The first term is caused by the carrier frequency and the second term is phase noise, and M is the ratio of the carrier frequency to STALO frequency. After reflected from a target, the received signal phase is that of the transmitted signal delayed by the round-trip time τ. The receiver output signal phase $\Phi(t)$, results from demodulating the received signal with the receiver STALO which has the same form as the transmitter STALO

$$\Phi_R(t) = 2\pi f_R t + M\varphi_R(t) \tag{3.58}$$

can be expressed as

$$\Phi(t) = 2\pi(f_R - f_T)t + 2\pi f_T \tau + M(\varphi_R(t) - \varphi_T(t - \tau)). \tag{3.59}$$

The first term is a frequency offset arising from non-identical STALO frequencies, which will bring a drift on the final image. Because this drift can be easily corrected using ground calibrator, it is ignored here. The second term forms the usual Doppler term which should be preserved. The last term represents the effect of STALO frequency instability which is of interest. It is assumed that $\varphi_T(t)$ and $\varphi_R(t)$ are independent random variables having identical power spectral density (PSD) $S_\varphi(f)$. Then the PSD of phase noise in a bistatic radar system is

$$S_{\varphi_B}(f) = 2M^2 S_\varphi(f) \tag{3.60}$$

where the factor 2 arises from the addition of two uncorrelated but identical PSDs. It has been proved that synchronization compensation is required for bistatic radar systems [54, 55].

Several potential synchronization techniques or algorithms [56–58], such as using ultra-high-quality oscillators [59] and an appropriate bidirectional link [60] are proposed. However, in most cases, we cannot alter the transmitter; hence, it is necessary to develop a practical synchronization technique without alteration to the transmitters. One potential solution is the following direct-path signal-based time and phase synchronization approach [61].

Suppose the nth transmitted pulse with carrier frequency f_{Tn} is

$$x_n(t) = s(t)\exp(j2\pi f_{Tn}t)\exp(j\varphi_{d(n)}) \tag{3.61}$$

where $\varphi_{d(n)}$ is the original phase, and $s(t)$ is the radar signal in baseband. Let t_{dn} denote the delay time of direct-path signal, the received direct-path signal is

$$s'_{dn}(t) = s(t - t_{dn})\exp[j2\pi(f_{Tn} + f_{dn})(t - t_{dn})]\exp(j\varphi_{d(n)}) \tag{3.62}$$

where f_{dn} is the Doppler frequency for the nth transmitted pulse. Suppose the demodulating signal in receiver is

$$S_f(t) = \exp{-j(2\pi f_{Rn}t)} \tag{3.63}$$

Hence the received signal in baseband is

$$s_{dn}(t) = s(t - t_{dn})\exp(-j2\pi(f_{Tn} + f_{dn})t_{dn})\exp(j2\pi\Delta f_n t)\exp(j\varphi_{d(n)}) \tag{3.64}$$

with $\Delta f_n = f_{Tn} - f_{Rn}$, here $\varphi_{d(n)}$ is the term to be extracted to compensate the phase synchronization errors in the reflected signals.

Matched filtering with the reference function yields

$$y_{dn}(t) \approx \Delta f_n \frac{\sin\left(\pi \Delta f_n\left(t - t_{dn} + \frac{\Delta f_n}{k_r}\right)\right)}{\pi \Delta f_n\left(t - t_{dn} + \frac{\Delta f_n}{k_r}\right)} \cdot \exp\left[j\pi \Delta f_n\left(t - t_{dn} + \frac{\Delta f_n}{k_r}\right)\right]$$

$$\cdot \exp\left\{-j\left[2\pi(f_{dn} + f_{Rn})t_{dn} - \frac{\pi \Delta f_n^2}{k_r} - \varphi_{d(n)}\right]\right\} \tag{3.65}$$

where k_r is the chirp rate of the transmitted signal. The maxima will be at $t = t_{dn} - \frac{\Delta f_n}{k_r}$ where we have

$$\exp\left[j\pi \Delta f_n\left(t - t_{dn} + \frac{\Delta f_n}{k_r}\right)\right]\Big|_{t=t_{dn}-\frac{\Delta f_n}{k_r}} = 1 \tag{3.66}$$

Hence the residual phase term in Eq. 3.65 is

$$\Psi(n) = -2\pi(f_{dn} + f_{Rn})t_{dn} - \frac{\Delta f_n^2}{k_r} + \varphi_{d(n)} \tag{3.67}$$

As Δf_n and k_r are typically on the orders of 1 kHz and 1×10^{13} Hz/s respectively [25], $\frac{\pi \Delta f_n^2}{\gamma}$ has negligible effects. Equation 3.67 can be simplified into

$$\psi(n) = -2\pi(f_{dn} + f_{Rn})t_{dn} + \varphi_{d(n)} \tag{3.68}$$

Similarly, we have

$$\psi(n+1) = -2\pi(f_{d(n+1)} + f_{R(n+1)})t_{d(n+1)} + \varphi_{d(n+1)} \tag{3.69}$$

Let

$$f_{d(n)} = f_{d0} + \delta f_{dn}, \quad f_{R(n)} = f_{R0} + \delta f_{Rn} \tag{3.70}$$

where f_{d0} and f_{R0} are the original Doppler frequency and error-free demodulating frequency in receiver, respectively. Correspondingly, δf_{dn} and δf_{Rn} are the frequency errors for the nth pulse. Hence we have

$$\varphi_{d(n+1)} - \varphi_{d(n)} = [\psi(n+1) - \psi(n)] - 2\pi(f_{R0} + f_{d0})(t_{d(n+1)} - t_{dn})$$

$$- 2\pi(\delta f_{dn} + \delta f_{Rn})(t_{d(n+1)} - t_{dn}) \tag{3.71}$$

Generally speaking, $\delta f_{dn} + \delta f_{Rn}$ and $t_{d(n+1)} - t_{dn}$ are typically on the orders of 10 Hz and 10^{-9}s, respectively, then $2\pi(\delta f_{dn} + \delta f_{Rn})(t_{d(n+1)} - t_{dn})$ is found to be smaller than $2\pi \times 10^{-8}$ radian, which has negligible effects. Furthermore, since $t_{d(n+1)}$ and t_{dn} are known, Eq. 3.71 is simplified into

$$\varphi_{d(n+1)} - \varphi_{d(n)} = \psi_e(n) \tag{3.72}$$

where

$$\psi_{e(n)} = [\psi(n+1) - \psi(n)] - 2\pi(f_{R0} + f_{d0})(t_{d(n+1)} - t_{dn}). \tag{3.73}$$

The $\varphi_{d(n)}$ can then be obtained by iterative computation of Eq. 3.73. Next, the phase synchronization errors in the reflected channel can be compensated with this information.

The direct-path signal-based synchronization approach provides a simple solution to compensate the synchronization errors, but there is a disadvantage in that the receiver must fly with a sufficient altitude and position to maintain a line-of-sight contact with the transmitter.

3.4.4 Challenges: Motion Compensation

In the previous discussions, we did not consider the motion errors. However, problems will arise due to the presence of atmospheric turbulence, which introduce platform trajectory deviations from normal position, as well as altitude (roll, pitch, and yaw angles) [62, 63]. Consequently, motion compensation is required. In current radar systems, GPS and inertial navigation systems (INS) are usually employed for this task. By contrast, for near-space vehicle-borne radars, motion compensation facilities may not be reachable because near-space vehicles have very limited load capability. Therefore, some efficient motion error compensation techniques should be developed.

For the near-space vehicle-borne BiSAR, since its synthetic aperture time is short, we ignore the acceleration errors in along-track and consider only the motion errors in cross-track in the following discussions. As shown in Fig. 3.15, suppose the ideal transmitter and receiver instantaneous positions at azimuth time τ_m are $(v_t\tau_m, y_{t0}, h_t)$ and $(v_r\tau_m, y_{r0}, h_r)$ respectively, and their actual positions are $(v_t\tau_m, y_{t0} + \Delta y_t(\tau_m), h_t)$ and $(v_r\tau_m, y_{r0}\Delta y_r(\tau_m), h_r)$.

Suppose the transmitter motion error in cross-track is $\Delta r_t(\tau_m)$, there are $\Delta y_t(\tau_m) = -\Delta r_t(\tau_m)\cos(\alpha_{t0})$ (α_{t0} is the instantaneous incidence angle from the transmitter to the point target $P_n(x_n, y_n, 0)$), $\Delta z_t(\tau_m) = -\Delta r_t(\tau_m)\sin(\alpha_{t0})$ and $y_n - y_{t0} = r_{tn}\cos(\alpha_{t0})$ with $r_{tn} = \sqrt{h_t^2 + (y_n - y_{t0})^2}$ and $h_t = r_{tn}\sin(\alpha_{t0})$. The transmitter range history can then be represented by

$$R_T(\tau_m) = \sqrt{(x_n - v_t\tau_m)^2 + (y_n - y_{t0} - \Delta y_t(\tau_m))^2 + (h_t - \Delta z_t(\tau_m))^2}$$
$$= \sqrt{(x_n - v_t\tau_m)^2 + r_{tn}^2 + 2r_{tn} \cdot \Delta r_t(\tau_m) + \Delta r_t^2(\tau_m)}. \tag{3.74}$$

Assume that the instantaneous transmitter squint angle is θ_{tm} and denote $x_t(\tau_m) = x_n - v_t\tau_m$ and $\tan(\theta_{tm}) = x_t(\tau_m)/r_{tn}$, we can then get

Fig. 3.15 Illustration of the
near-space vehicle-borne
BiSAR motion errors

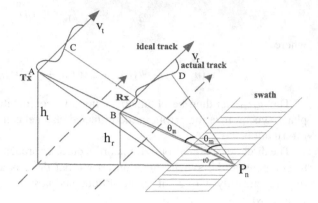

$$R_T(\tau_m) = \sqrt{r_{tn}^2 + x_t^2(\tau_m)} - \Delta r_t(\tau_m)\cos(\theta_{tm})$$

$$+ \sin(\theta_{tm})\sin(2\theta_{tm})\frac{\Delta r_t^2(\tau_m)}{2r_{tn}} + O\left(\frac{\Delta r_t(\tau_m)}{r_{tn}}\right)$$

$$\approx \sqrt{r_{tn}^2 + x_t^2(\tau_m)} - \Delta r_t(\tau_m)\cos(\theta_{tm}) \tag{3.75}$$

Similarly, for the receiver range history we have

$$R_R(\tau_m) \approx \sqrt{r_{rn}^2 + x_r^2(\tau_m)} - \Delta r_r(\tau_m)\cos(\theta_{rm}) \tag{3.76}$$

where r_{rn}, $x_r(\tau_m)$, $\Delta r_r(\tau_m)$ and θ_{rm} are defined in a manner similar to r_{tn}, $x_t(\tau_m)$, $\Delta r_t(\tau_m)$ and θ_{tm}. The bistatic range history can then be represented by

$$R(\tau_m) \approx \sqrt{r_{tn}^2 + x_t^2(\tau_m)} - \Delta r_t(\tau_m)\cos(\theta_{tm})$$

$$+ \sqrt{r_{rn}^2 + x_r^2(\tau_m)} - \Delta r_r(\tau_m)\cos(\theta_{rm}). \tag{3.77}$$

As the first and third terms are the ideal range history for subsequent image formation processing, we consider only the second and fourth terms. We have

$$\frac{\partial[\Delta r_t(\tau_m)\cos(\theta_{tm}) + \Delta r_r(\tau_m)\cos(\theta_{rm})]}{\partial \tau_m}$$

$$\doteq -\frac{\Delta r_t(\tau_{m1})}{r_{tn}}\sin(\theta_{tm})\cdot v_t \cdot (\tau_{m2} - \tau_{m1})$$

$$- \frac{\Delta r_r(\tau_{m1})}{r_{rn}}\sin(\theta_{rm})\cdot v_r \cdot (\tau_{m2} - \tau_{m1}) \tag{3.78}$$

As there are $\Delta r_t(\tau_{m1})/r_{tn} \ll 1$ and $\Delta r_r(\tau_{m1})/r_{rn} \ll 1$, when $\tau_{m2} - \tau_{m1}$ is short, Eq. 3.78 will be equal to zero. That is to say, the motion errors in a short interval can be seen as constant. This phenomenon validates again the equivalent velocity and NCS combined imaging algorithm. Otherwise, efficient motion compensation techniques should be applied. A potential solution is the subaperture-based motion compensation algorithms. The details can be found in [64, 65].

3.4.5 Challenges: Antenna Directing Synchronization

Another technological challenge is antenna directing synchronization, which requires the transmit antenna and receive antenna to simultaneously illuminate the same region on the ground. Although the feasibility of the BiSAR concept has already been demonstrated by experimental investigations [66], BiSAR antenna directing synchronization is still a technological challenge.

The concept of pulse chasing was proposed as a means for bistatic radar antenna directing synchronization [67, 68]. In this approach, the single receive beam rapidly scans the area in two dimensions covered by the transmit beam, essentially chasing the pulse as it propagates from the transmitter. Because the single beam chases one pulse at a time, this imposes a limit on the maximum allowable pulse repetition frequency (PRF). Moreover, antenna directing synchronization is even more demanding for BiSAR systems, especially for azimuth-variant BiSAR systems, since the geometric and radiometric characteristics of azimuth-variant BiSAR are strictly dictated by illuminator configuration and operation. The requirement of antenna directing synchronization for interferometric Cartwheel SAR was simply analyzed in [69]. Several altitude and antenna directing synchronization strategies for satellite formation configuration were proposed in [70]. The comparison of altitude and antenna pointing design strategies of noncooperative spaceborne BiSAR were investigated in [71].

Without loss of generality, we consider a rather general BiSAR configuration, in which the transmitter and receiver are mounted on different platforms. As a typical example, suppose the normalized transmit antenna gain is

$$G_T(\theta_T, \phi_T) = \exp\left\{-2\gamma\left[\left(\frac{\theta_T}{\theta_{T3\text{dB}}}\right)^2 + \left(\frac{\phi_T}{\phi_{T3\text{dB}}}\right)^2\right]\right\} \tag{3.79}$$

where γ is the Gauss parameter which is often assumed to be 1.3836, θ_T and ϕ_T are the antenna beamwidth in range and azimuth respectively. Correspondingly, θ_{T3dB} and ϕ_{T3dB} are their 3dB beamwidth. If there are range antenna directing errors $\Delta\theta_T$, we then have

$$\frac{\Delta G_T(\theta_T, \phi_T)}{G_T(\theta_T, \phi_T)} = -\frac{4\gamma\theta_T\Delta\theta_T}{\theta_{T3\text{dB}}^2} \tag{3.80}$$

Similarly, for the receive antenna we have

$$\frac{\Delta G_R(\theta_R, \phi_R)}{G_R(\theta_R, \phi_R)} = -\frac{4\gamma\theta_R\Delta\theta_R}{\theta_{R3\text{dB}}^2} \tag{3.81}$$

Then, there is [72]

$$\frac{\Delta A}{A} = \frac{1}{2}\left[\frac{\Delta G_T(\theta_T, \phi_T)}{G_T(\theta_T, \phi_T)} + \frac{\Delta G_R(\theta_R, \phi_R)}{G_R(\theta_R, \phi_R)}\right]$$

$$= -2\gamma\left[\frac{\theta_T}{\theta_{T3dB}^2}\Delta\theta_T + \frac{\theta_R}{\theta_{R3dB}^2}\Delta\theta_R\right]. \tag{3.82}$$

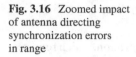

Fig. 3.16 Zoomed impact of antenna directing synchronization errors in range

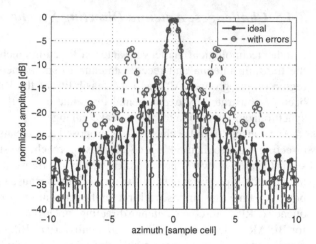

To evaluate the impact of antenna directing synchronization errors on SAR imaging performance, linear and quadratic errors are usually assumed in existent papers. In fact, antenna directing error usually is oscillatory for practical systems. Hence, we use an oscillatory model

$$\Delta\theta_T(t) = \Delta\theta_R(t) = A_r \cos(\omega_r t + \varphi_0) \tag{3.83}$$

with A_r, ω_r and φ_0 are the amplitude, frequency and initialization angle in range, respectively. Take an azimuth-invariant X-band BiSAR, in which the transmitter and receiver are moving in parallel tracks with a constant and identical velocities $v_s = 150$ m/s using the following parameters: $\theta_T(t) = 8°$, $\theta_R(t) = 5°$, $A_r = 1$, $\omega_r = 1.98$ and $\varphi_0 = 0$ as an example, the corresponding impacts are illustrated in Fig. 3.16. It is seen that antenna directing synchronization errors manifest themselves as a deterioration of the impulse response function. They may defocus BiSAR image and introduce a significant increase of the sidelobes.

Similarly, suppose the azimuth antenna directing synchronization errors are represented by

$$\Delta\beta_T = \Delta\beta_R = A_a \cos(\omega_a t + \beta_0) \tag{3.84}$$

where A_a, ω_a and β_0 are the amplitude, frequency and initialization angle in azimuth, respectively. The corresponding transmit/receive azimuth antenna figure can be represented by

$$w(t) = w_a[t - A_a \sin(\omega_a t + \beta_0)] \tag{3.85}$$

with $w_a(t) = \text{sinc}^2(\frac{\pi L_a v_s}{\lambda R_c}(t - t_c))$, where L_a, R_c and t_c are the antenna length, nearest range from the platform to the scene and its corresponding time, respectively.

From the SAR processing procedure, we know that the azimuth signal after range and azimuth compressions is

$$S_{out}(t) = \exp(j2\pi f_c t - j\pi k_r t^2) \cdot \int_{-T_s/2}^{T_s/2} w(\tau)\exp(2\pi k_r t\tau)d\tau \qquad (3.86)$$

Since

$$t \gg A_a \sin(\omega_a t + \beta_0) \qquad (3.87)$$

According to the principle of Taylor series expansion, we have

$$w(t) = w_a(t) - A_a \sin(\omega_a t + \beta_0) \cdot w_a'(t) \qquad (3.88)$$

Substitute Eq. 3.88 to Eq. 3.86 we get [73]

$$S_{out}(t) = \exp(j2\pi f_c t - j\pi k_r t^2) \cdot \int_{-T_s/2}^{T_s/2} w_a(\tau) \exp(2\pi k_r t\tau)d\tau$$

$$- \exp(j2\pi f_c t - j\pi k_r t^2) \cdot A_a \int_{-T_s/2}^{T_s/2} \sin(\omega_a \tau + \beta_0) \exp(2\pi k_r t\tau)\, d\tau.$$

$$(3.89)$$

We can see that, similar to range antenna directing synchronization errors, azimuth antenna directing synchronization errors also introduce paired echoes.

Thus, a high-precision antenna directing synchronization approach should be developed for near-space vehicle-borne azimuth-variant BiSAR systems.

3.5 Conclusion

In this chapter, we considered the near-space vehicle-borne passive radar in surveillance and reconnaissance. It has the advantages of entire planet coverage and simple transmitter-receiver synchronization, but high-resolution performance cannot be obtained. Thus, we further considered near-space vehicle-borne BiSAR high-resolution imaging. A comprehensive description of the system imaging performance was provided, along with the signal processing challenges. Also, we considered near-space vehicles in a passive environment monitoring. Potential applications in homeland security and disaster monitoring and technical challenges in synchronization compensation and motion compensation are investigated.

References

1. Moccia, A., Salzillo, G., D'Errico, M., Rufino, G., Alberti, G.: Performace of spaceborne bistatic synthetic aperture radar. IEEE Trans. Aerosp. Electron. Syst. **41**, 1383–1395 (2005)
2. Li, X.R., Jilkov, V.P.: Survey of maneuvering target tracking, part II: motion models of ballistic and space targets. IEEE Trans. Aerosp. Electron. Syst. **46**, 196–119 (2010)

3. Allen, E.H.: The case for near-space. Aerosp. Am. **22**, 31–34 (2006)
4. Tomme, E.B.: The paradigm shift of effects-based space: near-space as a combat space effects enabler. http://www.airpower.au.af.mil (2009). Accessed May 2010
5. Progri, I.: Geolocation of RF Signals: Principles and Simulations. Springer, London (2011)
6. Marcel, M.J., Baker, J.: Interdisciplinary design of a near-space vehicle. In: Proceedings of Southeast Conference, Richmond, VA, 421–426 (2007)
7. Guan, M.X., Guo, Q., Li, L.: A novel access protocol for communication system in near-space. In: Proceedings of Wireless Communication and Network Mobile Computation Conference, Shanghai, China, 1849–1852 (2007)
8. Wang, W.Q., Cai, J.Y., Peng, Q.C.: Near-space SAR: a revolutionay microwave remote sensing mission. In: Proceedings of Asia-Pacific Synthetic Aperture Radar Conference, Huangshan, China, 127–131 (2007)
9. Galletti, M., Krieger, G., Thomas, B., Marquart, M., Johannes, S.S.: Concept design of a near-space radar for tsunami detection. In: Proceedings of IEEE Geoscience Remote Sensors Symposium, Barcelona, 34–37 (2007)
10. Wang, W.Q., Cai, J.Y., Peng, Q.C.: Near-space microwave radar remote sensing: potential and challenge analysis. Remote Sens. **2**, 717–739 (2010)
11. Wang, W.Q.: Application of near-space passive radar for homeland security. Sens. Imag: Int. J. **8**, 39–52 (2007)
12. Zavorotny, V.U., Voronovich, A.G.: Scattering of GPS signals from the ocean with wind remote sensing applications. IEEE Trans. Geosci. Remote Sens. **38**, 951–964 (2000)
13. Heise, S., Wickert, J., Beyerle, G., Schmidt, T., Smit, H., Cammas, J.P., Rothacher, M.: Comparison of water vapor and temperature results from GPS radio occultation aboard CHAMP with MOZAIC aircraft measurements. IEEE Trans. Geosci. Remote Sens. **46**, 3406–3411 (2008)
14. Garrison, J.L., Komjathy, A., Zavorotny, V., Katzberg, S.J.: Wind speed measurements using forward scattered GPS signals. IEEE Trans. Geosci. Remote Sens. **40**, 50–65 (2002)
15. Gleason, S., Hodgart, S., Sun, Y.P, Gommenginger, C., Mackin, S., Adjrad, M., Unwin, M.: Wind speed measurements using forward scattered GPS signals. IEEE Trans. Geosci. Remote Sens. **43**, 1229–1241 (2005)
16. Bindlish, R., Crow, W.T., Jackson, T.J.: Role of passive microwave remote sensing in improving flood forecasts. IEEE Geosci. Remote Sens. Lett. **6**, 112–116 (2009)
17. Cherniakov, M.: Bistatic Radar: Emerging Technology. Wiley, New York (2007)
18. Wang, W.Q., Cai, J.Y.: A technique for jamming bi- and multi-static SAR systems. IEEE Geosci. Remote Sens. Lett. **4**, 80–82 (2007)
19. Wang, W.Q.: Multi-Antenna Synthetic Aperture Radar Imaging: Principles and Applications (in Chinese). National Defense Industry Press, Beijing (2011)
20. Grewal, M.S., Weill, L.R., Andrews, A.P.: Global Positioning Systems: Inertial Navigation and Integration. Wiley, New York (2001)
21. He, F., Cherniakov, M., Zeng, T.: Signal detectability in SS-BSAR with GNSS non-cooperative transmitter. IEE Proc. Radar Sonar Navig. **152**, 124–132 (2005)
22. Gleason, S., Hodgart, S., Sun, Y., Gommenginger, C., Mackin, S., Adjrac, M., Unwin, M.: Detection and processing of bistatically reflected GPS signals from low earth orbit for the purpose of ocean remote sensing. IEEE Trans. Geosci. Remote Sens. **43**, 1229–1241 (2005)
23. Cherniakov, M., Saini, R., Antoniou, M., Zuo, R., Plakidis, E.: Experiences gained during the development of a passive BSAR with GNSS transmitters of opportunity. Int. J. Navig. Observ. **1**, 1–12 (2008)
24. Wang, W.Q.: Near-space passive radar for homeland security: potential and challenge. In: Proceedings of XXI International Society Photogrammetry Remote Sensors Symposium, Beijing, China, 1021–1027 (2008)
25. Krieger, G., Moccia, A.: Spaceborne bi- and multistatic SAR: potential and challenges. IEE Proc. Radar Sonar Navig. **153**, 184–198 (2006)
26. Neo, Y.L., Wong, F.H., Cumming, I.G.: Processing of azimuth-invariant bistatic SAR data using the range Doppler algorithm. IEEE Trans. Geosci. Remote Sens. **46**, 14–21 (2006)

27. Nico, G., Tesauro, M.: On the existence of coverage and integration time regimes in bistatic SAR configurations. IEEE Geosci. Remote Sens. Lett. **4**, 426–430 (2007)
28. Wang, W.Q., Cai, J.Y.: Azimuth-variant bistatic synthetic aperture radar data processing. In: Daniels, J.A. (ed.) Advances in Environmental Research. NOVA Publisher, New York (2011)
29. Marcos, J.S., Dekker, P.L., Mallorqui, J.J., Aguasca, A., Prats, P.: SABRINA: a SAR bistatic receiver for interferometric applications. IEEE Geosci. Remote Sens. Lett. **4**, 307–311 (2007)
30. Liebe, J.R., van de Giesen, N., Andreini, M.S., Steenhuis, T.S., Walter, M.T.: Suitability and limitations of ENVISAT ASAR for monitoring small reservoirs in a semiarid area. IEEE Trans. Geosci. Remote Sens. **47**, 1536–1547 (2009)
31. Wong, F.H., Yeo, T.S.: New application of nonlinear chirp scaling in SAR data processing. IEEE Trans. Geosci. Remote Sens. **39**, 946–953 (2001)
32. Loffeld, O., Nies, H., Peters, V., Knedlik, S.: Models and useful relations for bistatic SAR processing. IEEE Trans. Geosci. Remote Sens. **42**, 2031–2038 (2004)
33. Neo, Y.L., Wong, F.H., Cumming, I.G.: A two-dimensional spectrum for bistatic SAR processing using series reversion. IEEE Geosci. Remote Sens. Lett. **4**, 93–96 (2007)
34. Wong, F.H., Neo, Y.L., Cumming, I.G.: Focusing bistatic SAR data using the nonlinear chirp scaling algorithm. IEEE Trans. Geosci. Remote Sens. **46**, 2493–2505 (2008)
35. Geng, X.P., Yan, H.H., Wang, Y.F.: A two-dimensional spectrum model for general bistatic SAR. IEEE Trans. Geosci. Remote Sens. **46**, 2216–2223 (2008)
36. Liou, Y.A., Pavelyev, A.G., Liu, A.G., Pavelyev, A.A., Yen, N., Huang, C.Y., Fong, C.J.: FORMOSAT-3/COSMIC GPS radio occultation mission: preliminary results. IEEE Trans. Geosci. Remote Sens. **45**, 3813–3826 (2007)
37. Kursinski, E.R., Hajj, G.A., Schofield, J.T.: Observing earth's atmosphere with occultation measurements using the global positioning system. J. Geophys. Rev. **102**, 23429–23465 (1997)
38. Lantilhac, S.: UAV flight plan optimized for sensor requirements. IEEE Aerosp. Electron. Syst. Mag. **25**, 11–14 (2010)
39. Joseph, A.T., Vander Velde, R., O'Neill, P.E., Lang, R.H., Gish, T.: Soil moisture retrieval during a corn growth cycle using L-band (1.6GHz) radar observations. IEEE Trans. Geosci. Remote Sens. **46**, 2365–2374 (2008)
40. Renaux, A., Atallah, N.L., Forster, P., Larzabal, P.: A useful form of the Abel bound and its application to estimator threshold prediction. IEEE Trans. Sig. Process **55**, 2365–2369 (2007)
41. Eigel, R., Collins, P., Terzuoli, T., Nesti, G., Fortuny, J.: Bistatic scattering characterization of complex objects. IEEE Trans. Geosci. Remote Sens. **38**, 2078–2092 (2000)
42. Wang, W.Q.: Conceptual design of near-space radar for ocean remote sensing. In: Proceedings of International Workshop Advances SAR Oceanography from ENVISAT and ERS Missions, Italy, 1–5 (2008)
43. Wang, W.Q., Cai, J.Y., Peng, Q.C.: Passive ocean remote sensing by near-space vehicle-borne GPS receiver. In: Tang, D.L. (ed.) Remote Sensing of the Changing Oceans, Springer-Verlag, Heidelberg (2011)
44. Kouchi, K., Yamazaki, F.: Characteristics of tsunami-affected areas in moderate-resolution satellite images. IEEE Trans. Geosci. Remote Sens. **45**, 1650–1657 (2007)
45. Tralli, D.M., Blom, R.G., Zlotnichi, V., Donnellan, A., Evans, D.L.: Satellite remote sensing of earthquake, volcano, flood, landslide and coastal inundation hazards. ISPRS J. Photogramme Remote Sens. **59**, 185–198 (2005)
46. Bovolo, F., Bruzzone, L.: A split-based approach to unsupervised change detection in large-size multitemporal images: application to tsunami-damage assessment. IEEE Trans. Geosci. Remote Sens. **45**, 1658–1679 (2007)
47. Meyers, R.G., Draim, C.J.E., Cefola, P.J., Raizer, V.Y.: A new tsunami detection concept using space-based microwave radiometry. In: Proceedings of IEEE Geoscience Remote sensors Symposium, Boston, MA, 958–961 (2008)
48. Borrero, J.C.: Field data and satellite imagery of tsunami effects in Banda Aceh. Science **308**, 1596–1597 (2005)

49. Meyer, F., Hinz, S., Laika, A., Weihing, D., Bamler, R.: Performance analysis of the TerraSAR-X traffic monitoring concept. ISPRS J. Photogramm. Remote Sens. **6**, 225–242 (2006)
50. Le Caillec, J.M.: Study of the SAR signature of internal waves by nonlinear parametric autoregressive models. IEEE Trans. Geosci. Remote Sens. **44**, 148–158 (2006)
51. Rodenas, J.A., Garello, R.: Internal wave detection and location in SAR images using wavelet transform. IEEE Trans. Geosci. Remote Sens. **36**, 1494–1507 (1998)
52. Hogan, G.G., Chapman, R.D., Watson, G., Thompson, D.R.: Observations of ship-generated interal waves in SAR images. IEEE Trans. Geosci. Remote. Sens. **34**, 532–542 (1996)
53. Auterman, J.L.: Phase stability requirements for a bistatic SAR. In: Proceedings of IEEE Naturalist Radar Conference, Atlanta, Georgia, 48–52 (1984)
54. Wang, W.Q.: Analytical modeling and simulation of phase noise in bistatic synthetic aperture radar systems. Fluct. Noise. Lett. **6**, 297–303 (2006)
55. Wang, W.Q.: Clock timing jitter analysis and compensation for bistatic synthetic aperture radar systems. Fluct. Noise Lett. **7**, 341–350 (2007)
56. Wang, W.Q.: Approach of adaptive synchronization for bistatic SAR real-time imaging. IEEE Trans. Geosci. Remote Sens. **45**, 2695–2700 (2007)
57. Wang, W.Q.: GPS-based time & phase synchronization processing for distributed SAR. IEEE Trans. Aerosp. Electron Syst. **45**, 1040–1051 (2009)
58. Wang, W.Q.: Bistatic synthetic aperture radar synchronization processing. In: Kouemou, G. (ed.) Radar Technology. In-Tech Press, India (2010)
59. Gierull, C.: Mitigation of phase noise in bistatic SAR systems with extremely large synthetic apertures. In: Proceedings of Europe Synthetic Aperture Radar Symposium, Dresden, Germany, 1251–1254 (2006)
60. Younis, M., Metzig, R., Krieger, G.: Performance prediction of a phase synchronization link for bistatic SAR. IEEE Geosci. Remote Sens. Lett. **3**, 429–433 (2006)
61. Wang, W.Q., Ding, C.B., Liang, X.D.: Time and phase synchronization via direct-path signal for bistatic synthetic aperture radar systems. IET Radar Sonar Navig. **2**, 1–11 (2008)
62. Dickey, F.M., Doerry, A.W., Romero, L.A.: Degrading effects of the lower atmosphere on long range airborne synthetic aperture radar imaging. IET Radar Sonar Navig. **1**, 329–339 (2007)
63. Fornaro, G., Franceschetti, G., Pema, S.: Motion compensation errors: effects on the accuracy of airborne SAR images. IEEE Trans. Aerosp. Electron Syst. **41**, 1338–1351 (2005)
64. Carrara, W.G., Goodman, R.S., Majewski, R.M.: Spotlight Synthetic Aperture Radar: Signal Processing Algorithm. Artech House, Norwood (1995)
65. Potsis, A., Reigber, A., Mittermayer, J., Moreira, A., Uzunoglou, N.: Sub-aperture algorithm for motion compensation improvement in wide-beam SAR data processing. Electron. Lett. **37**, 1405–1407 (2001)
66. Wendler, M., Krieger, G., Horn, R.: Results of a joint bistatic airborne SAR experiment. In: Proceedings of International Radar Symposium, Dresden, Germany, 247–253 (2003)
67. Jackson, M.C.: The geometry of bistatic radar systems. IEE Proc. Pt. F **133**, 604–612 (1986)
68. Schoenenberger, J.G., Forrest, J.R.: Principles of independent receivers for use with cooperative radar transmitters. Radio Electron. Eng. **52**, 93–101 (1982)
69. Massonnet, D.: Capabilities and limitations of the interferometric cartwheel. IEEE Trans. Geosci. Remote Sens. **39**, 506–520 (2001)
70. D'Errico, M., Mocccia, A.: Altitude and antenna pointing design of bistatic radar formations. IEEE Trans. Aerosp. Electron. Syst. **39**, 949–959 (2003)
71. Huang, H.F., Liang, D.N.: The comparison of altitudeand antenna pointing design strategies of noncooperative spaceborne bistatic radar. USA, 568–571 (2005)
72. Tang, Z.Y., Zhang, S.R.: Bistatic Synthetic Aperture Radar System and Principle (in Chinese). National Defense Press, Beijing, China (2003)
73. Wang, W.Q., Cai, J.Y.: Antenna directing synchronization for bistatic synthetic aperture radar systems. IEEE Antenna Wireless Propag. Lett. **9**, 307–310 (2009)

Chapter 4
Near-Space Vehicles in High-Resolution Wide-Swath Remote Sensing

Abstract Spaceborne SAR has an imaging capability of wide-swath (the width of the ground area covered by the radar beam) with a limited azimuth resolution. By contrast, airborne SAR has an imaging capability of high azimuth resolution, but limited swath coverage. There is, therefore, a desire to increase swath coverage and azimuth resolution simultaneously. As near-space vehicles operate at altitudes higher than that of airplanes but lower than satellites with a high flying speed, compared to spaceborne and airborne SARs, simultaneous relative high-resolution and wide-swath (HRWS) remote sensing is possible for near-space vehicle-borne SAR. In this chapter, we explained how near-space vehicles could be exploited for future HRWS remote sensing applications.

Keywords Near-space · High-resolution wide-swath · Multiple apertures · Waveform diversity · Multiple-input and multiple-output (MIMO) · Space-time coding

4.1 Restrictions on Achievable Resolution and Swath

An efficient remote sensing technique should provide high resolution imagery over a wide area of surveillance, although there is a contradiction between azimuth resolution and swath width [1, 2]. A good azimuth resolution requires a short antenna to illuminate a long synthetic aperture which results in a wide Doppler bandwidth. This calls for a high pulse repeated frequency (PRF) to sample the Doppler spectrum. Thus, azimuth resolution and ambiguity suppression impose a lower bound on the PRF, and the higher it is, the better the achievable azimuth resolution becomes and the smaller the ambiguous signals will be. But, a high PRF means a smaller swath width. The relationship between the maximum imaging swath on ground, W_s, and the required PRF can be expressed as

$$W_s \leq \frac{c_0}{2 \cdot \sin(\eta) \cdot PRF} \tag{4.1}$$

where c_0 and η denote the speed of light and the incidence angle, respectively.

W.-Q. Wang, *Near-Space Remote Sensing*,
SpringerBriefs in Electrical and Computer Engineering,
DOI: 10.1007/978-3-642-22188-0_4, © Wen-Qin Wang 2011

Substituting the expression of SAR azimuth resolution ρ_α and rearranging the terms in Eq. 4.1, we then have

$$\frac{W_s}{\rho_\alpha} \leq \frac{c_0}{2 \cdot v_s \cdot \sin(\eta)} \tag{4.2}$$

where v_s is the SAR platform velocity. Generally, c_0/v_s is nearly constant at 20,000 for LEO spaceborne SARs and typically in the range of 300,000–750,000 for airborne SARs. Near-space vehicles can fly at a speed ranging from stationary to 1,500 m/s, the corresponding c_0/v_s will be greater than 100,000. Thus, compared to spaceborne and airborne SARs, the near-space vehicle-borne SAR provides a more flexible choice between azimuth resolution and swath width for HRWS remote sensing.

Equation 4.2 can also be reformed into the basic minimum antenna area constraint [3]

$$A_{\text{antenna}} = H_a \cdot L_a \geq \frac{4 v_s \lambda R_c \cdot \tan(\eta)}{c_0} \tag{4.3}$$

where H_a is the antenna width, L_a is the antenna length, λ is the radar wavelength, and R_c is the slant range from the radar to the mid-swath. This requirement arises because the illuminated area of the ground must be restricted so that the radar does not receive ambiguous returns in range dimension or/and azimuth dimension. In this respect, a high operating PRF is desirable to suppress azimuth ambiguity. But the magnitude of the operating PRF is limited by range ambiguity requirement. Therefore, in conventional SARs the unambiguous swath width and the achievable azimuth resolution impose contradicting requirements on system design, and consequently allows only for a concession between azimuth resolution and swath width. These considerations can be combined into one inequality which expresses the range of slant range values that are appropriate to each of the unambiguous swath intervals

$$\left(\frac{n_a}{PRF} + T_p\right) \frac{c_0}{2} \leq R_c \leq \left(\frac{n_a + 1}{PRF} - T_p\right) \frac{c_0}{2} \tag{4.4}$$

where n_a is an integer that effectively labels each swath interval, and T_p is the pulse duration.

Although near-space vehicle-borne SAR provides a more flexible choice between azimuth resolution and swath width, the illuminated swath must be restricted so that the received radar echoes are unambiguous in range dimension or/and azimuth dimension. As an example, assuming an X-band near-space vehicle-borne SAR system with the following typical parameters: $\lambda = 0.03\,\text{m}$, $\eta = 60°$, $v_s = 1,000\,\text{m/s}$, $L_a = 0.4\,\text{m}$, $H_a = 0.22\,\text{m}$, and the flying altitude is $H_0 = 60\,\text{km}$ (according to [4], $H_0 = 60\,\text{km}$ is the best altitude for high-speed near-space vehicles), then the maximal swath width is found to be 32.7 km.

The attainment of a wider swath will become increasingly difficult if a higher azimuth resolution is required, due to the requirements of increased PRF. The PRF limit for a certain swath W_s and a given incidence angle η is [5]

$$\frac{2v_s}{L_a} \leq PRF \leq 0.8 \frac{c_0}{2W_s \sin(\eta)} \tag{4.5}$$

4.2 State-of-the-Art: HRWS Remote Sensing

Numerous innovative concepts are presented to enable HRWS remote sensing whereof the most promising ones employ a single transmit antenna in combination with multiple channels on receive [5–11]. This means that these systems have a receive antenna which is split into multiple subapertures with independent receiver chains that are interpreted as individual channels. A similar technique is the distribution of the receiver apertures on multiple platforms that are grouped into a formation of space or airborne sensors leading to a multistatic SAR [12–15]. It is common to all of these methods that the backscattered signals are received simultaneously by multiple apertures, which are mutually displaced in azimuth and/or elevation dimension. The different receive aperture positions introduce spatial diversity in the received echoes. The basic idea is, hence, to use the multiple receivers to gather additional information and to benefit from this information to overcome the above restrictions of conventional SARs [16–23]. In general, the multiple receivers can be either arranged in flight direction ("along-track"), perpendicular to it ("cross-track"), or in both dimensions. They are illustrated respectively in Fig. 4.1.

4.2.1 Multiple Apertures in Elevation

The multiple apertures in elevation proposed by Griffith and Mancini [16] consists of an array antenna split in elevation. The overall antenna dimension is smaller than that implied by the minimum antenna area constraint yielding a broad beam in elevation dimension that covers a wide swath but at the same time gives rise to range ambiguous echoes. The range ambiguities are suppressed by adaptively steering nulls in the antenna pattern in elevation to the directions of the ambiguous returns. In this way, a widened swath can be obtained. However, in a monostatic SAR the swath will be no longer continuous because blind ranges are introduced when the receiver is switched off during transmission. This restriction can be overcome by using bistatic SAR which allows for simultaneous transmission and reception [24–26].

In 2001, Suess et al. [5] proposed an innovative processing approach in elevation dimension for HRWS remote sensing. The SAR system is built on an array antenna consisting of multiple elements in elevation. The processing concept combines the echoes from the subapertures in elevation in a way to form a narrow beam in elevation which "scans" the ground in real-time in order to follow the echo of the transmitted

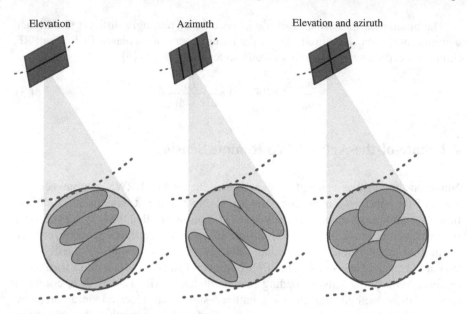

Fig. 4.1 Typical configurations of the multiple apertures or receivers

signal on ground. This enables the suppression of range-ambiguous returns and ensures a high antenna gain. However, this technique requires knowledge of the observed terrain topography, or else a mispointing of the narrow elevation beam may occur, resulting in severe gain loss [27].

4.2.2 Multiple Channels in Azimuth

In 1992, Currie and Brown [2] proposed the displaced phase center antenna (DPCA) in azimuth technique, which is based on dividing the receive antenna in along-track direction into multiple subapertures, each receiving, down-converting, and digitizing the radar echoes. Hence, for every transmitted pulse the system receives multiple pulses in the along-track direction. This means that additional samples can be gathered, thus increasing the effective sampling rate on receive. Consequently, either the resolution can be improved while the swath width remains constant, or the swath can be widened without increasing azimuth ambiguities or impairing the resolution. That is to say, the system benefits from the whole antenna length regarding azimuth ambiguity suppression, while azimuth resolution is determined by the dimension of a single subaperture, thus decoupling the restrictions on HRWS remote sensing. However, this approach imposes a stringent timing requirement on the system regarding the relation between sensor velocity, PRF, and antenna length.

Regarding multiple channels in azimuth, the approach presented in [28, 29] introduces a phase correction that is applied to the raw data to resample the signal in azimuth. This method is based on an analysis of the multichannel signal's phase compared to the phase of a monostatic and uniformly sampled signal. This yields a Doppler frequency-dependent phase difference between the multichannel signals and the monostatic signal. Hence, by applying an appropriate phase correction to the data of each individual channel, the phase of the multichannel signal is adjusted in such a way that the resulting phase corresponds to the monostatic and uniformly sampled signal. In 2003, Krieger et al. [30] proposed an algorithm based on a generalization of the sampling theorem that allows for unambiguous recovery of the azimuth spectrum from multiple aliased subaperture signals. This method is elaborated in several follow-on papers [31–36] and extended to the burst operation mode [37].

4.2.3 Multiple Apertures in Two Dimensions

In 1999, Callaghan and Longstaff [17] proposed the quad array approach which is based on an antenna split into two rows and two columns yielding a four-element array. This approach can be understood as a combination of the two approaches described previously. Thus, this approach combines the advantages of gathering additional samples in azimuth to suppress azimuth ambiguities and simultaneously enabling an enlarged swath for a fixed PRF. However, the proposed system in elevation will also result in blind ranges in the imaged swath. Additionally, the stringent timing constraint to ensure a uniform spatial distribution of the gathered sampled in azimuth dimension must also be satisfied.

Besides, Claassen and Eckerman [38] proposed an alternative concept based on steering multiple beams to different azimuth directions and assigning each of the corresponding footprints to a different slant range. In other words, the footprint of each beam steered to a different squint angle corresponds to a different subswath of the overall imaged region. As the antenna elevation dimension is larger than the imaged swath, range ambiguities can be well suppressed, thus allowing for a PRF high enough to suppress the azimuth ambiguities. However, there will be coarsened resolution and impaired performance arising from the needed high squint angles [22].

4.2.4 Distributed SAR Constellations

The concept of sparse array SAR constellations was proposed in [12, 13, 23]. This approach considers an arbitrarily distributed spaceborne array of radar satellites that do not necessarily build a formation aligned in the along-track dimension. The satellites are then regarded as a sparse antenna array that collect also angle-of-arrival information. This additional information enables wide-swath imaging with enhanced azimuth resolution. However, the processing of the irregularly spaced and sparsely distributed samples of the array requires a space-time minimum mean-squared error estimator. An optimum way of processing in the space-time domain is derived

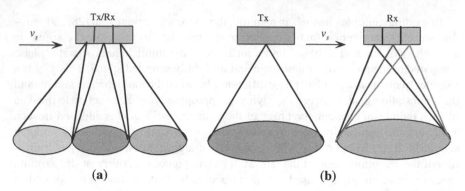

Fig. 4.2 Geometry mode of the near-space vehicle-borne SAR for HRWS remote sensing

in [13]. Further, Aguttes [14] proposed the concept of SAR train consisting of a multi-satellite constellation distributed in along-track direction and uses a spread spectrum waveform for transmission.

4.3 Near-Space Vehicle-Borne SAR HRWS Remote Sensing

To reduce the required PRF for near-space vehicle-borne SAR HRWS remote sensing, we consider first the multiple azimuth beam technique. The incentive is that it allows an effective decrease in the operating PRF of a SAR system, while ensuring that the bandwidth of target signals is still adequately sampled in azimuth. Two different approaches (see Fig. 4.2) to implementing this technique are described below.

4.3.1 Single-Phase Center Multibeam SAR Imaging

The operation mode of the single phase center multibeam (SPCM) SAR system is to transmit pulses by a single broad azimuth beam and receive the returns by a number of narrow contiguous azimuth beams that span the mainlobe width of the transmit beam, as shown in Fig. 4.3. A distinct channel is associated with each of the receive beams, and hence, the data are split according to the azimuth angular position or, equivalently, instantaneous Doppler frequency center in the azimuth direction. As a result, given the knowledge of the relative squint angles of each beam (hence the Doppler center frequency for each beam) and assuming suitable isolation between the beams, each channel can be sampled at a Nyquist rate appropriate to the bandwidth covered by each narrow beam, instead of that covered by the full beamwidth.

This arrangement enables correct sampling of the azimuth spectrum with a PRF fitting the total antenna azimuth length, which is N times smaller than the PRF necessary for the antenna azimuth length, L_a. The area of each beam antenna is then restricted by

Fig. 4.3 Geometry mode of the near-space vehicle-borne SAR for HRWS remote sensing. (Reproduced by permission of © 2009 Elsevier Masson SAS)

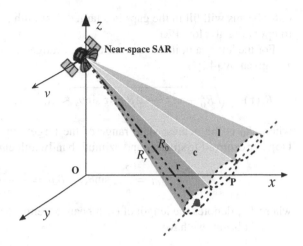

$$A_{as} \geq \frac{4v_s \lambda R_c \tan(\eta)}{c_0} \cdot \frac{1}{N} \tag{4.6}$$

Clearly, the minimum area of each beam antenna is N-times smaller than the respective area of a monostatic SAR. Correspondingly, the relationship expressed in Eq. 4.2 is changed into

$$\frac{W_s}{\rho_\alpha} \leq \frac{N c_0}{2 \cdot v_s \cdot \sin(\eta)} \tag{4.7}$$

From Eq. 4.7 we note that the relation not only depends on the platform velocity v_s and the incidence angle η but also on the number of antenna beams.

Thereafter, the DPCA technique discussed in [39, 40] can be used to gain additional samples along the synthetic aperture which enables an efficient suppression of the azimuth ambiguities, i.e., the multiple beams in azimuth allow for the division of a broad Doppler spectrum into multiple narrow-band subspectra with different Doppler centroids. A coherent combination of the subspectra will then yield a broad Doppler spectrum for high azimuth resolution. Thus this technique is especially attractive for high-resolution SAR that uses a long antenna for unambiguous wide-swath coverage.

The data in each channel will be aliased to the relative zero-frequency. This problem can be resolved by using the prior knowledge of the relevant Doppler center frequencies to regain the full azimuth Doppler bandwidth. Consider Fig. 4.3 [41] which shows a near-space vehicle-borne SAR with three beams in azimuth. For a given coordinate of the SAR, the imaged area is simultaneously illuminated by the three beams. We may view this SAR as a conventional SAR (the central beam), operating with a PRF which is one third of that required to adequately sample its beamwidth, together with two additional beams on either side of the central one. As described previously, the basic idea is that the additional samples obtained by the outer beams will fill in the gaps in a target's azimuth phase history which occur due to operating at a low PRF.

For the left beam, its instantaneous slant range in terms of the slow time variable τ is given by [42]

$$R_l(\tau) = \sqrt{R_0^2 + v_s^2 \tau^2 - 2R_0 v_s \tau \sin\theta_s} \approx R_0 - v_s \tau \sin\theta_s + \frac{v_s^2 \tau^2 \cos^2\theta_s}{2R_0} \quad (4.8)$$

where R_0 is the nearest slant range of the target, and θ_s is the squint angle. The Doppler centroid frequency and azimuth bandwidth can be derived from Eq. 4.8 as

$$f_{dl} = \frac{2}{\lambda} v_s \sin\theta_s, \quad B_{dl} = \frac{2v_s \cos\theta_s}{L_{as}} \quad (4.9)$$

where L_{as} denotes the length of each beam antenna. Similarly, for the central beam and right beam, we have

$$f_{dc} = 0, \quad B_{dc} = \frac{2v_s}{L_{as}} \quad (4.10)$$

$$f_{dr} = -\frac{2}{\lambda} v_s \sin\theta_s, \quad B_{dr} = \frac{2v_s \cos\theta_s}{L_{as}} \quad (4.11)$$

Generally, we have the following approximations because the θ_s is small.

$$B_{dl} \approx B_{dc} \approx B_{dl} = B_{ds}, \quad f_{dl} - f_{dc} = f_{dc} - f_{dr} = B_{ds} \quad (4.12)$$

The ambiguous Doppler spectrum can be recovered unambiguously by applying a system of reconstruction filters. Some reconstruction algorithms have been proposed by other authors, e.g., [34, 31]. A block diagram for the reconstruction from the three-channel signals is shown in Fig. 4.4. This algorithm is based on considering the multichannel near-space vehicle-borne SAR data acquisition as a linear system with multiple channels, each described by a linear filter. From the sampling theorem, we know that the sampled signal spectrum $X_s(f)$ is the sum of the unsampled signal spectrum, $X_0(f)$, it repeats every f_s Hz with f_s the sampling frequency.

There is

$$X_s(f) = f_s \sum_{-\infty}^{\infty} X_0(f - nf_s) \quad (4.13)$$

If $f_s \geq 2B_{ds}$, the replicated spectra will not be overlapped, and the original spectrum can be regenerated by chopping $X_s(f)$ off above $f_s/2$. Thus, $X_0(f)$ can be reproduced from $X_s(f)$ through an ideal low-pass filter that has a cutoff frequency of $f_s/2$. The subsequent data processing involves interpolating the data from each channel. We then have

$$y_s(n) = \begin{cases} x_s\left(\frac{n}{N_0}\right), & n = 0, \pm N_0, \pm 2N_0, \cdots \\ 0, & \text{otherwise} \end{cases} \quad (4.14)$$

Fig. 4.4 Multichannel
signal reconstruction
algorithm in case of three
channels

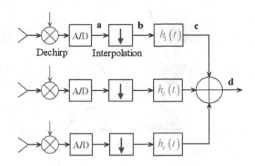

where N_0 is the interpolation scaling factor. Accordingly, the spectrum is represented by

$$Y_s(f) = \sum_{n=-\infty}^{\infty} y_s(n)e^{-j2\pi fn} = X_s(N_0 f) \qquad (4.15)$$

Then, the linear filter is derived as

$$H_k(f) = \begin{cases} N_0, & \left| f - \frac{f_{ck}}{N_0} \right| \le \frac{B_{ds}}{N_0}, k \in (l, c, r) \\ 0, & \text{otherwise} \end{cases} \qquad (4.16)$$

where $k \in (l, c, r)$ is shown in Fig. 4.4, and f_{ck} is the corresponding Doppler centroid.

Finally, the filtered signals can be combined coherently, as shown in Fig. 4.5. That is to say, in a manner similar to adaptive antenna beamforming technique, the filtered signals can be combined coherently. In this way, the capability of ambiguity suppression allowing for improved resolution and an enlarged swath can be achieved. Note that, for optimum performance the relationship between sensor velocity and along-track offsets of the three subchannels must result in equally spaced effective phase centers, so that a uniform sampling of the received signal can be obtained [31].

4.3.2 Multiple Phase Center Multibeam SAR Imaging

The multiple phase center multibeam (MPCM) SAR system also synthesizes multiple receive beams in the azimuth direction; however, the operating mode of this system is quite different from that of the previous one. In this case, the system transmits a single broad beam and receives the radar returns in multiple beams which are displaced in the along-track direction. The motivation is that multiple independent sets of target returns are obtained for each transmitted pulse if the distance between

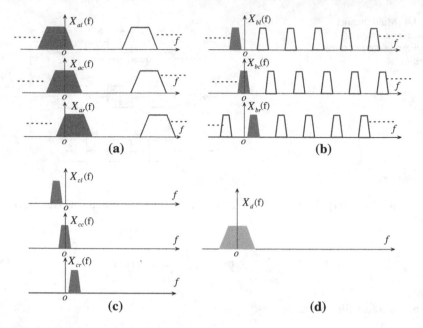

Fig. 4.5 Azimuthal spectra synthesis for multichannel subsampling in case of three channels. (Reproduced by permission of © 2009 Elsevier Masson SAS)

phase centers is suitably set. This method basically implies that we may broaden the azimuth beam from the diffraction-limited width, giving rise to improved resolution, without having to increase the system operating PRF.

4.3.2.1 Imaging Scheme and Signal Model

Consider Fig. 4.3, for the central beam there is

$$R_c(\tau) = \sqrt{R_0^2 + v_s^2(\tau - \tau_0)^2 - 2R_0 v_s(\tau - \tau_0)\sin\theta_s}$$

$$\approx R_0 - v_s \sin\theta_s(\tau - \tau_0) + \frac{v_s^2 \cos^2\theta_s}{2R_0}(\tau - \tau_0)^2 \qquad (4.17)$$

where τ_0 is the azimuth reference time. The phase history can then be represented by

$$\Phi_c(\tau) = \frac{4\pi R_c(\tau)}{\lambda} \qquad (4.18)$$

Suppose the spatial separation between two phase centers is d_a, the phase histories of the left beam and right beam can be represented, respectively, by

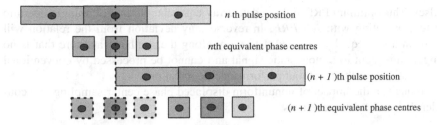

Fig. 4.6 Equivalent phase centers for the case of three receive beams

$$\Phi_l(\tau) = \frac{2\pi R_c(\tau)}{\lambda} + \frac{2\pi R_c \left(\tau - \frac{d_a}{v_s}\right)}{\lambda} \tag{4.19}$$

$$\Phi_r(\tau) = \frac{2\pi R_c(\tau)}{\lambda} + \frac{2\pi R_c \left(\tau + \frac{d_a}{v_s}\right)}{\lambda}. \tag{4.20}$$

From the DPCA principle, we known that subsequent data processing simply involves interleaving the data from each stream properly in the azimuth direction so that the proper relative phasing is maintained. Azimuth compression can then be processed as normal. This technique may be extended to more than three receive beams, the main effect being that the constant phase difference increases in magnitude for the additional beams. Note that the application of this technique assumes that cubic (and high order) terms may be neglected in the azimuth phase history of targets. A general criterion for this condition to hold is that this term causes less than a $\pi/2$ excursion in phase over the aperture synthesis time.

To construct the synthetic aperture, the system must operate with a PRF which leads to a properly sampled synthetic aperture appropriate to the beamwidth of the system. In fact, the operating PRF of the system is always equal to the Nyquist rate for the diffraction-limited beamwidth of the antenna, regardless of the number of receive beams and the width of the beams. This is illustrated in Fig. 4.6 for the case of three receive beams.

4.3.2.2 Non-uniform Displaced Phase Center Sampling

Taking into account that the overall antenna length in azimuth L_a is made up of N apertures, each length is d_a, the optimum PRF can be derived from

$$PRF_{uni} = \frac{2v_s}{Nd_a} = \frac{2v_s}{L_a} \tag{4.21}$$

This imposes a stringent requirement on the system as it states that to ensure equal spacing between all samples in azimuth the PRF has to be chosen such that the SAR platform moves just one half of its antenna length L_a between subsequent radar pulses. This optimum PRF yields a data array equivalent to that of a single-aperture system operating with $N \cdot PRF$. In reverse, any deviation from the relation will result in a non-equally sampled data array along the synthetic aperture that is no longer equivalent to a monostatic signal and cannot be processed by conventional monostatic algorithms without performance degradation.

To analyze the impact of nonuniform displaced phase center sampling, we consider the received radar returns

$$s_i(t, \tau) \approx \sigma_i[h_0(t) \otimes_t h_{1,i}(t, \tau)], \quad i = 1, 2, \ldots, N \qquad (4.22)$$

where σ_i is the RCS parameter, t is the range fast time, τ is the azimuth slow time, and $\otimes_t$ is a convolution operator on the variable t. $h_0(t)$ and $h_{1,i}(t, \tau)$ denote, respectively, the range reference function and azimuth reference function

$$h_0(t) = w_r(t) \cdot \exp\left(-j\pi k_r t^2\right) \qquad (4.23)$$

$$h_{1,i}(t, \tau) = \exp\left\{-j\frac{2\pi}{\lambda}\left[R_c(\tau) + R_c\left(\tau + i\frac{d_a}{v_s}\right)\right]\right\}$$
$$\times w_a(\tau) \cdot \delta\left[\tau - \frac{R_c(\tau) + R_c\left(\tau + i\frac{d_a}{v_s}\right)}{c_0}\right] \qquad (4.24)$$

where $w_r(t)$ and $w_a(\tau)$ denote the antenna pattern in range dimension and azimuth dimension, respectively. As there is

$$R_c(\tau) + R_c\left(\tau + i\frac{d_a}{v_s}\right) \approx 2R_c\left(\tau + i\frac{d_a}{2v_s}\right) \qquad (4.25)$$

we then have

$$s_i(t, \tau) \approx \sigma_i\left[h_0(t) \otimes_t h_1\left(t, \tau + i\frac{d_a}{2v_s}\right)\right] \qquad (4.26)$$

with

$$h_1\left(t, \tau + i\frac{d_a}{2v_s}\right) = w_a\left(\tau + i\frac{d_a}{2v_s}\right) \cdot \exp\left\{-j\frac{4\pi}{\lambda}R_c\left(\tau + i\frac{d_a}{2v_s}\right)\right\}$$
$$\cdot \delta\left[\tau - \frac{2R_c\left(\tau + i\frac{d_a}{2v_s}\right)}{c_0}\right] \qquad (4.27)$$

Equivalently, the nonuniform PRF can be considered as azimuth time drift.

$$\tau_{er} = \frac{d_a}{2v_s} - \frac{1}{N \cdot PRF} \qquad (4.28)$$

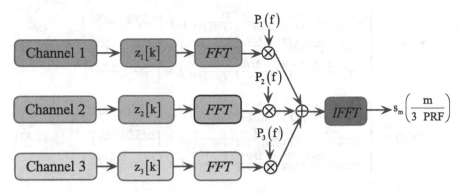

Fig. 4.7 Reconstruction filtering for multichannel subsampling in case of three channels

After matched filtering and range mitigation correction, we get

$$
s_i \left(k \frac{1}{N \cdot PRF} \right) = w_a \left(k \frac{1}{N \cdot PRF} + i \cdot \tau_{er} \right)
$$
$$
\times \exp \left\{ -j \left[2\pi f_d \left(\frac{1}{N \cdot PRF} + i \cdot \tau_{er} \right) \right. \right.
$$
$$
\left. \left. + \pi k_a \left(k \frac{1}{N \cdot PRF} + i \cdot \tau_{er} \right)^2 \right] \right\} \qquad (4.29)
$$

where k is an integer, f_d is the Doppler frequency centroid, and k_a is the Doppler chirp rate. It is noted that the signals are periodic nonuniform with the period of $1/PRF$. This information is particularly important for developing nonuniform reconstruction algorithms.

4.3.2.3 Azimuth Signal Reconstruction Processing

The ambiguous Doppler spectrum of a nonuniformly sampled SAR signal can be recovered unambiguously by applying a system of reconstruction filters. The algorithm illustrated in Fig. 4.7 is based on considering the data acquisition in a DPCA SAR as a linear system with multiple receiver channels, each described by a linear filter. The reconstruction consists essentially of multiple linear filters which are individually applied to the subsampled signals of the receiver channels and then combined coherently.

Taking three azimuth beams as an example, the corresponding reconstruction filters are

$$
P_1(f) = \begin{cases} \dfrac{2j \sin(j2\pi T \cdot PRF)}{j(4\sin(2\pi T \cdot PRF) - 2\sin(4\pi T \cdot PRF))}, & f \in \left[\frac{-3PRF}{2}, \frac{-PRF}{2} \right] \\[2mm] \dfrac{-2j \sin(j4\pi T \cdot PRF)}{j(4\sin(2\pi T \cdot PRF) - 2\sin(4\pi T \cdot PRF))}, & f \in \left[\frac{-PRF}{2}, \frac{PRF}{2} \right] \\[2mm] \dfrac{-2j \sin(j4\pi T \cdot PRF)}{j(4\sin(2\pi T \cdot PRF) - 2\sin(4\pi T \cdot PRF))}, & f \in \left[\frac{PRF}{2}, \frac{3PRF}{2} \right] \end{cases} \qquad (4.30)
$$

$$P_2(f) = \begin{cases} \dfrac{-K \cdot \exp(j2\pi(f+PRF)T))}{j(4\sin(2\pi T \cdot PRF) - 2\sin(4\pi T \cdot PRF))}, & f \in \left[\dfrac{-3PRF}{2}, \dfrac{-PRF}{2}\right] \\[2ex] \dfrac{G \cdot \exp(j2\pi(f-PRF)T)}{j(4\sin(2\pi T \cdot PRF) - 2\sin(4\pi T \cdot PRF))}, & f \in \left[\dfrac{-PRF}{2}, \dfrac{PRF}{2}\right] \\[2ex] \dfrac{-K \cdot \exp(j2\pi(f-2PRF)T)}{j(4\sin(2\pi T \cdot PRF) - 2\sin(4\pi T \cdot PRF))}, & f \in \left[\dfrac{PRF_p}{2}, \dfrac{3PRF_p}{2}\right] \end{cases} \quad (4.31)$$

$$P_3(f) = \begin{cases} \dfrac{K' \cdot \exp(j2\pi(f+PRF)T))}{j(4\sin(2\pi T \cdot PRF) - 2\sin(4\pi T \cdot PRF))}, & f \in \left[\dfrac{-3PRF}{2}, \dfrac{-PRF}{2}\right] \\[2ex] \dfrac{-G' \cdot \exp(j2\pi(f-PRF)T)}{j(4\sin(2\pi T \cdot PRF) - 2\sin(4\pi T \cdot PRF))}, & f \in \left[\dfrac{-PRF}{2}, \dfrac{PRF}{2}\right] \\[2ex] \dfrac{K' \cdot \exp(j2\pi(f-2PRF)T)}{j(4\sin(2\pi T \cdot PRF) - 2\sin(4\pi T \cdot PRF))}, & f \in \left[\dfrac{PRF}{2}, \dfrac{3T \cdot PRF}{2}\right] \end{cases} \quad (4.32)$$

where $T = d_a/(2v_s)$, $K = \exp(j2\pi T \cdot PRF) - 1$, and $G = \exp(j4\pi T \cdot PRF) - 1$. The reconstructed signals can be expressed as

$$s_a\left(\frac{m}{3PRF}\right) = z_1\left(\frac{m}{3PRF}\right) \otimes p_1\left(\frac{m}{3PRF}\right) + z_2\left(\frac{m}{3PRF}\right) \otimes p_2\left(\frac{m}{3PRF}\right)$$
$$+ z_3\left(\frac{m}{3PRF}\right) \otimes p_3\left(\frac{m}{3PRF}\right) \quad (4.33)$$

with

$$z_1(k) = \begin{cases} s_m\left(\frac{k}{3PRF}\right) & k = 3n \\ 0 & k \neq 3n \end{cases} \quad (4.34)$$

$$z_2(k) = \begin{cases} s_m\left(\frac{k}{3PRF} - \frac{d_a}{2v_s}\right) & k = 3n \\ 0 & k \neq 3n \end{cases} \quad (4.35)$$

$$z_3(k) = \begin{cases} s_m\left(\frac{k}{3PRF} + \frac{d_a}{2v_s}\right) & k = 3n \\ 0 & k \neq 3n \end{cases} \quad (4.36)$$

where $p_1(\cdot)$, $p_2(\cdot)$ and $p_3(\cdot)$ are the time-domains representing $P_1(f)$, $P_2(f)$ and $P_3(f)$, respectively.

4.3.3 Ambiguity-to-Signal Ratio Analysis

For a given range and azimuth antenna pattern, the PRF must be selected such that the total ambiguity noise contribution is adequately small relative to the signal, i.e., from -18 to $-20\,dB$. A low PRF will increase the azimuth ambiguity level due to increased aliasing of the azimuth spectra. On the other hand, a high PRF value will reduce the interpulse period and result in overlap between the received pulses in time. To resolve these problems, the transmit interference restriction on the PRF must be satisfied with

$$\frac{n'}{\frac{2R_{\min}}{c_0} - T_p} < PRF < \frac{n'+1}{\frac{2R_{\max}}{c_0}} \tag{4.37}$$

where $R_{\min}$ is the nearest slant range, $R_{\max}$ is the farthest slant range, n' is a given integer, and T_p is the pulse duration.

Similarly, the nadir interference restriction on the PRF can be represented by

$$\frac{m'}{\frac{2R_{\min}}{c_0} - 2T_p - \frac{2h_s}{c_0}} < PRF < \frac{m'}{\frac{2R_{\max}}{c} - \frac{2h_s}{c_0}} \tag{4.38}$$

where m' is a given integer and h_s is the platform altitude. Alternatively, given a PRF or range of PRFs, the antenna dimensions and/or weighting (to lower the sidelobe energy) must be adequately small such that the ambiguity-to-signal ratio (ASR) specification is satisfied.

The azimuth ambiguities arise from finite sampling of the Doppler spectrum at intervals of the PRF. Since azimuth spectrum repeats at the PRF intervals, the signal components outside this frequency interval will fold back into the main part of the spectrum, and the desired signal band will be contaminated by the ambiguous signals from adjacent spectra. This can be evaluated by the azimuth ambiguity to signal ratio (AASR) defined as [43]

$$AASR \approx \frac{\sum_{\substack{m=-\infty \\ m \neq 0}}^{\infty} \int_{-0.5B_d}^{0.5B_d} G^2(f + m \cdot PRF)\, df}{\int_{-0.5B_d}^{0.5B_d} G^2(f)\, df} \tag{4.39}$$

where B_d and $G(f)$ denote the SAR correlator's azimuth processing bandwidth and azimuth antenna pattern, respectively.

Considering the three beams illustrated in Fig. 4.3, we have

$$G_k(\theta) = \mathrm{sinc}^2 \left(\frac{\pi L_{as} \cos(i \cdot \theta_s)}{\lambda} \sin(\theta - i \cdot \theta_s) \right), \quad i \in (-1, 0, 1), k \in (l, c, r) \tag{4.40}$$

Note that $i \in (-1, 0, 1)$ is determined from the positions of the three beams, i.e., $i = -1$ is for the left beam, $i = 0$ is for the central beam and $i = 1$ is for the right beam. Because the 3 dB beamwidth is approximately determined by [43]

$$\theta \approx \frac{\lambda}{2v_s} f, \quad \theta_s = k_a \frac{\lambda}{L_{as}} \tag{4.41}$$

with k_a a given constant. Equation 4.40 can then be further simplified into

$$G_k(f) \approx \mathrm{sinc}^2 \left[\pi L_{as} \cos \left(i \cdot \frac{k_a c_0}{f L_{as}} \right) \left(\frac{f}{2v_s} - i \cdot \frac{k_a}{L_{as}} \right) \right] \tag{4.42}$$

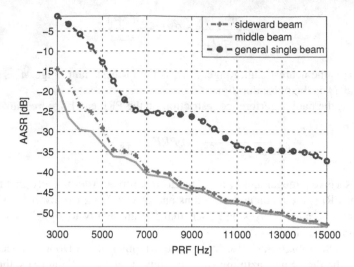

Fig. 4.8 The AASR of an example near-space vehicle-borne SAR as a function of PRF. (Reproduced by permission of © 2009 Elsevier Masson SAS)

Then, from Eq. 4.39 we can get

$$
AASR_k(PRF) = \left\{ \sum_{\substack{m=-\infty \\ m\neq 0}}^{\infty} \left[\int_{(i-0.5)B_{ds}}^{(i+0.5)B_{ds}} G_k^2(f + m \cdot PRF) \, df \right. \right.
$$

$$
\left. \left. + \sum_{j\neq k} \int_{(i-0.5)B_{ds}}^{(i+0.5)B_{ds}} G_k(f + m \cdot PRF) G_j(f + m \cdot PRF) \, df \right] \right\}
$$

$$
\cdot \left\{ \int_{(i-0.5)B_{ds}}^{(i+0.5)B_{ds}} G_k^2(f) \, df + \sum_{j\neq k} \int_{(i-0.5)B_{ds}}^{(i+0.5)B_{ds}} G_k(f) G_j(f) \, df \right\}^{-1}, \quad j \in (l, c, r).
$$

$$(4.43)$$

An example near-space wide-swath SAR is considered with the following parameters: $h_s = 60$ km, $v_s = 1,000$ m/s, $\lambda = 0.03$ m, $L_{as} = 1.2$ m, the calculated AASR results are illustrated in Fig. 4.8. Note that the AASR is typically specified to be on the order of -20 dB; however, even at this value ambiguous signals may be observed in images that have very bright targets adjacent to dark targets, because SAR imagery can have an extremely wide dynamic range due to the correlation compression gain. As such, a lower AASR, e.g., -30 dB, is desirable.

Similarly, range ambiguities may result from the preceding and succeeding pulse echoes arriving at the antenna simultaneously with the desired return. This type of ambiguity is relatively not significant for near-space SAR. The range ambiguity of the near-space vehicle-borne HRWS SAR can be analyzed similar to the general SAR, which has been fully investigated, e.g., [3].

Thus the set of PRFs is established by the acceptable maximum range and AASR requirement, as well as the transmit and nadir interference. Note that, there may be no acceptable PRFs at some incidence angles that meet the minimum requirements. The designers then have the option to relax the performance specifications for the imaging area or exclude these modes from the operations plan.

4.3.4 Conceptual System Design

An example system is designed to manifest the imaging performance. It operates in X-band with a center frequency of 10 GHz. From the radar equation, we obtain [24]

$$NESZ = \frac{8\pi R_s^3 v_s \lambda K T_{sys} F_n L_f}{P_{avg} N A_{as}^2 \rho_r} \tag{4.44}$$

where R_s is the average slant range which is assumed constant, K ($K \approx 1.38 \times 10^{-23}$) is the Boltzmann constant, T_{sys} is the system noise temperature, L_f is the loss factor, F_n is the receiver noise figure, P_{avg} is the average transmit power, N is also the number of subapertures, A_{as} is the sub-aperture antenna area, and ρ_r is the range resolution cell for one look.

To calculate the system performance we assume a range resolution of $\rho_r = 0.2\,m$, an overall loss factor of $L_f = 3$ dB, and a receiver noise figure of $F_n = 3$ dB. It is further assumed that the signal bandwidth is adjusted for varying incidence angles such that the ground-range resolution is constant across the whole swath. An example system design is provided in Table 4.1. We note that, for the incidence angle given in Table 4.1, the swath width is about 8 km and the NESZ is approximately −48 dB. These results show that a satisfactory performance can be achieved, however, with only a small number of subapertures which have relatively small antenna area. The conclusion is that unambiguous range and swath width can be obtained using a near-space vehicle-borne SAR with multiapertures in azimuth.

4.4 Near-Space HRWS Remote Sensing via Multiple Apertures

The DPCA technique allows either the azimuth resolution to be improved while the swath width remains constant, or the PRF to be reduced without increasing the azimuth ambiguities. However, it imposes a stringent timing requirement on the system regarding the relationship among platform velocity, PRF, and antenna length; otherwise, the azimuth signal will be nonuniformly sampled. As digital beam forming (DBF) on receive is a promising candidate for HRWS remote sensing [44–47], in this section we describe one reflector antenna-based DBF solution to near-space vehicle-borne SAR HRWS remote sensing.

Table 4.1 Performance parameters of an example near-space vehicle-borne SAR

Parameters	Variables	Values
Mean transmit power	P_{avg}	1 W
Number of sub-aperture	N	3
Sub-aperture antenna length	L_{as}	0.40 m
Sub-aperture antenna width	H_a	0.10 m
Near-space vehicle velocity	v_s	500 m/s
Incidence angle	η	20°
Swath width	W_s	6.80 km
Radiometric resolution	NESZ	−48.58 dB
Incidence angle	η	25°
Swath width	W_s	7.30 km
Radiometric resolution	NESZ	−48.11 dB
Incidence angle	η	35°
Swath width	W_s	8.94 km
Radiometric resolution	NESZ	−46.73 dB

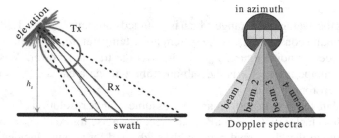

Fig. 4.9 Reflector antenna based system architecture with DBF in elevation and azimuth

4.4.1 System Architecture and Imaging Scheme

In SAR remote sensing missions the antenna is a key element in the total system performance. Designing one customized antenna with features such as wide swath and high resolution is a technical challenge in current SAR systems. As SAR antenna acts like a two-dimensional spatial filter, requirements imposed on both azimuth pattern and elevation pattern are important. To obtain HRWS imaging for the near-space vehicle-borne SAR, a reflector antenna with multiple feed elements and digital beamforming technique are employed. As shown in Fig. 4.9, the reflector antenna based near-space vehicle-borne SAR involves digital beamforming in elevation and digital beamforming in azimuth which are discussed, respectively, in the following two subsections.

4.4.1.1 Digital Beamforming in Elevation

To implement wide-swath imaging, the reflector antenna based digital beamforming in elevation is employed. The reflector antenna consists of a parabolic reflector and a feed array of transmit/receive elements. To illuminate a given angular segment in elevation, only the corresponding feed elements are activated. On transmit activating all elements generate a wide beam illuminating the complete swath. On receive the reflected signals illuminate the entire reflector antenna, but they are focused on individual feed elements. The receive beam scans the complete swath within the time period of one pulse repetition interval $1/PRF$, whereas each element is only active during a subinterval of this time period. When a high PRF is employed, multiple portions of the swath will be illuminated instantaneously but each actives a different subset of the feed elements because of different angle-of-arrival [48, 49]. Equivalently, a wide swath can be obtained.

For a strictly spherical Earth model, the angle-of-arrival of a point target echo is associated by

$$\eta_s(t_d) = \arccos\left(\frac{(h_s + R_e)^2 + R^2_e + R^2(t_d)}{2(h_s + R_e)R(t_d)}\right) \tag{4.45}$$

where R_e is the Earth radius and $R(t_d)$ with t_d the two-way time delay is the slant-range distance to one given target. Thus, there is a one-to-one relation between the required beam steering angle $\eta_s(t_d)$ and the time variable t_d. This information provides a potential solution to wide-swath SAR imaging. It is well known that, to avoid range ambiguities resulting from the preceding and succeeding pulse echoes arriving at the antenna simultaneously, the slant range R should be [50]

$$\frac{c_0}{2}\left(\frac{i}{PRF} + T_p + \Delta T_{tr}\right) < R < \frac{c_0}{2}\left(\frac{i+1}{PRF} - T_p\right) \tag{4.46}$$

where i is an integer, T_p is the pulse duration, and ΔT_{tr} is the switching time between the transmission and reception of a pulse.

Consider a SAR which operates with a PRF appropriate to the desired azimuth resolution requirement, but with a wider (M times) swath than that implied by Eq. 4.46, we then have

$$\frac{c_0}{2}\left(\frac{i}{PRF} + T_p + \Delta T_{tr}\right) < R < \frac{c_0}{2}\left(\frac{i+M}{PRF} - T_p\right) \tag{4.47}$$

As the idea of scan-on-receive (SCORE) based digital beamforming technique is to shape a time varying elevation beam in reception such that it follows the echo of the pulse on the ground [5], we divide the whole swath into M sub swaths.

$$\frac{c_0}{2}\left(\frac{i+m-1}{PRF} + T_p + \Delta T_{tr}\right) < R_m < \frac{c_0}{2}\left(\frac{i+m}{PRF} - T_p\right), 1 \leq m \leq M \tag{4.48}$$

As a compromise between computation complexity and imaging performance, here M is determined by

$$M = \lceil\frac{2R_{\max} \cdot PRF}{c_0}\rceil - \lfloor\frac{2R_{\min} \cdot PRF}{c_0}\rfloor \tag{4.49}$$

where $\lceil \cdot \rceil$ and $\lfloor \cdot \rfloor$ denote respectively the maximum and minimum integers, R_{max} and R_{min} denote the maximum and minimum slant ranges within the imaged swath.

To avoid range ambiguity, for the conventional SAR system the PRF should be satisfied with

$$\frac{n'}{\frac{R_{min}}{c_0} - T_p - \tau_{rp}} < PRF < \frac{n'+1}{\frac{R_{max}}{c_0} + \tau_{rp}} \tag{4.50}$$

where n' is a given integer, and τ_{tp} is the receiver protecting window extension about the pulse duration, T_p. However, by applying the SCORE operation mode, for the reflector antenna-based SAR system the PRF can only be satisfied with the relation

$$\frac{n'}{\frac{R_m}{c_0} - T_p - \tau_{rp}} < PRF < \frac{n'+1}{\frac{R_{m+1}}{c_0} + \tau_{rp}} \tag{4.51}$$

where R_m and R_{m+1} are the slant ranges to two adjacent subswaths, respectively. Comparing Eqs. 4.50 and 4.51, we can see that, for the same operating PRF a higher Doppler bandwidth can be sampled allowing for an improved azimuth resolution while keeping the range ambiguities constant. Alternately, the PRF can be reduced without an increase of azimuth ambiguities and degradation of the azimuth resolution while increasing the unambiguous imaging swath width.

Suppose the looking-down angles of the first formed subaperture to each subswath are $\alpha_1(r), \alpha_2(r), \ldots, \alpha_M(r)$, the relative phase delay from the first subswath to each formed subaperture can then be represented by [51]

$$0, \frac{2\pi d_r \sin(\alpha_1(r))}{\lambda}, \ldots, \frac{2\pi (M-1)d_r \sin(\alpha_1(r))}{\lambda} \tag{4.52}$$

where d_r denotes the elevation distance between two subapertures. Similarly, for the second subswath we have

$$0, \frac{2\pi d_r \sin(\alpha_2(r))}{\lambda}, \ldots, \frac{2\pi (M-1)d_r \sin(\alpha_2(r))}{\lambda} \tag{4.53}$$

Similar relations can be obtained for the remaining subswaths. They can be formed as a matrix expressed in Eq. 4.54,

$$A_r(r) = \begin{bmatrix} 1 & 1 & \cdots & 1 \\ \exp\left(j\frac{2\pi d_r \sin(\alpha_1)}{\lambda}\right) & \exp\left(j\frac{2\pi d_r \sin(\alpha_2)}{\lambda}\right) & \cdots & \exp\left(j\frac{2\pi d_r \sin(\alpha_M)}{\lambda}\right) \\ \vdots & \vdots & \ddots & \vdots \\ \exp\left(j\frac{2\pi (M-1)d_r \sin(\alpha_1)}{\lambda}\right) & \exp\left(j\frac{2\pi (M-1)d_r \sin(\alpha_2)}{\lambda}\right) & \cdots & \exp\left(j\frac{2\pi (M-1)d_r \sin(\alpha_M)}{\lambda}\right) \end{bmatrix} \tag{4.54}$$

where

$$\alpha_m = \alpha \left(\frac{c_0}{2}\left(t + \frac{i+m-1}{PRF}\right)\right), \quad 1 \leq m \leq M \tag{4.55}$$

with

$$\alpha(x) = \arccos \left[\frac{x^2 + h_s^2 + 2h_s R_e}{2x(h_s + R_e)} \right] - \eta \qquad (4.56)$$

where t represents the sampling time (fast time). As an analog-to-digital convertor (ADC) is placed after each T/R-module in the feed array, a posteriori, digital beamforming on receive can then be formed in the direction of a wanted subswath. Equivalently, a large swath can be synthesized.

4.4.1.2 Digital Beamforming in Azimuth

To further alleviate the requirements of HRWS imaging imposed on the minimum antenna area, digital beamforming in azimuth is further employed. In planar antenna-based digital beamforming in azimuth systems, all subapertures cover the same angular segment, thus "seeing" the identical Doppler spectra. Consequently, for the same PRF a higher (e.g., N times) Doppler bandwidth can be sampled allowing for an improved azimuth resolution while keeping the range ambiguities constant. Alternatively, the PRF can be reduced by $1/N$ while increasing the unambiguous range by a factor of N thus extending the equivalent imaging swath. However, from the sampling theorem, we know that the sampled signal spectrum is the sum of the unsampled signal spectrum.

As the Doppler spectra is undersampled, the Doppler spectrum can only be recovered unambiguously through the combination of the total spatial samples. That is to say, subsequent azimuth processing must combine the total N receive channels, each subsampled with PRF and aliased in frequency domain, to a single channel of $N \cdot PRF$ that is free of aliasing. Moreover, the relationship between platform velocity and along-track offsets of the azimuth subchannels must result in equally spaced effective phase centers, otherwise, there will be nonuniform spatial sampling which makes the subsequent Doppler spectrum synthesis a challenge.

In contrast to conventional planar antenna-based systems, for the reflector antenna-based digital beamforming in azimuth systems, there are single transmit feed and multiple receive feeds that are displaced in the along-track direction. Each azimuth element illuminates at a different angle and covers a distinct angular segment. Thus, each element samples a narrow Doppler spectrum. To exploit the large antenna array for signal transmission, the multidimensional waveform encoding investigated in [52] is modified and applied in azimuth direction. As shown in Fig. 4.10, in azimuth a series of subpulses instead of a wide duration pulse are transmitted, each subpulse is separately transmitted with a different short time delay by different transmit beams.

Note that the PRF must be high enough such that the spatial sampling of each beam or channel is adequate. If the Doppler spectra of the elements are contiguous, they can jointly yield a higher azimuth resolution, $D/(2N)$, where D is the reflector antenna diameter.

Fig. 4.10 Modified
multidimensional waveform
encoding in azimuth

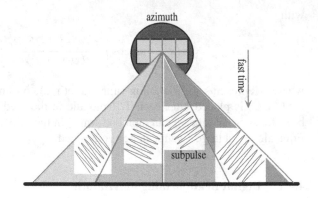

4.4.2 Imaging Performance Analysis

As discussed previously, the antenna area will affect the azimuth resolution and the available swath width. Additionally, antenna beam shape, especially its sidelobe characteristics, is also key to the imaging performance. Ambiguity noise is thus an important consideration. The constraints expressed in Eqs. 4.46 and 4.47 are approximate and the required signal to ambiguity noise ratio may not be met. Thus, it is necessary to analyze the RASR and AASR performances.

4.4.2.1 RASR Performance Analysis

At a given time t within the data record window, range ambiguous signals arrive from the ranges of

$$R_j = \frac{c_0}{2} \left(\frac{m+j-1}{PRF} + t \right), m = 1, 2, \ldots, M; j = \pm 1, \pm 2, \ldots, \pm N_h \quad (4.57)$$

where j, the pulse number ($j = 0$ for the desired pulse), is positive for the preceding pulses and negative for the succeeding ones. $j = N_h$ is the number of pulses to the horizon. We can see that this expression is different from the expression for single-aperture SAR systems, which is

$$R_j = \frac{c_0}{2} \left(\frac{j}{PRF} + t \right), j = \pm 1, \pm 2, \ldots, \pm N_h. \quad (4.58)$$

The RASR is determined by summing all signal components within the data record window arising from the preceding and succeeding pulses, and taking the ratio of this sum to the integrated signal return from the desired pulse

$$RASR = \frac{\sum_j \int_{R_{min}}^{R_{max}} \sum_1^M \frac{\gamma_{jm} G_j^2 \sigma_j}{R_j^3 \sin(\alpha_j)}}{\int_{R_{min}}^{R_{max}} \frac{G_0^2 \sigma_0}{R_0^3 \sin(\alpha_0)}} \tag{4.59}$$

where γ_{jm} is the beamforming gain of the mth subswath at the range of R_j, G_j is the cross-track antenna pattern at the jth time interval of the data recording window at a given α_j, σ_j is the corresponding normalized backscatter coefficient. The G_0, σ_0, R_0 and α_0 are the corresponding parameters of the desired unambiguous return.

$$AASR_k(PRF) = \left\{ \left[\sum_{\substack{m=-\infty \\ m \neq 0}}^{\infty} \left[\int_{f_{dc,k}-B_d/2}^{f_{dc,k}+B_d/2} G_k^2(f+m \cdot PRF)\,df \right. \right. \right.$$

$$\left. + \sum_{j \neq k} \int_{f_{dc,k}-B_d/2}^{f_{dc,k}+B_d/2} \Gamma_{k,j} \cdot G_k(f+m \cdot PRF) \cdot G_j(f+m \cdot PRF)\,df \right] \right]$$

$$\times \left\{ \int_{f_{dc,k}-B_d/2}^{f_{dc,k}+B_d/2} G_k^2(f)\,df + \sum_{j \neq k} \int_{f_{dc,k}-B_d/2}^{f_{dc,k}+B_d/2} \Gamma_{k,j} \cdot G_k(f) \cdot G_j(f)\,df \right\}^{-1}$$

$$\tag{4.60}$$

4.4.2.2 AASR Performance Analysis

The desired azimuth signal will also be contaminated by the ambiguous signals coming from adjacent spectra. For the reflector antenna-based SAR system, the AASR for the kth beam can be derived as Eq. 4.60 [43], where $(k, j) \in [1, 2, 3, \dots,]$ and

$$G_{k/j}(f) = \text{sin c}^2 \left(\frac{L_a \cdot \cos(\eta_{k/j}) \cdot (f - f_{dc,k/j})}{2v_s} \right) \tag{4.61}$$

is the azimuth antenna pattern, L_a is the azimuth antenna length, $\eta_{k/j}$ and $f_{dc,k/j}$ are the squint angle and Doppler frequency centroid for the k or jth channel, respectively. As the antenna steering angle is computed using the correspondence between the signal delay and angle of arrival, any steering error will result in a SCORE loss. An antenna pattern loss factor ($\Gamma k, j$) is thus employed in Eq. 4.60, which can be calculated as [48]

$$\Gamma_{k,j} = \frac{\int_{(j-k)\tau_d-T_{ps}/2}^{(j-k)\tau_d+T_{ps}/2} |G_r\left(\tau - \frac{2R_0}{c_0}\right)|^2 d\tau}{\int_{-T_{ps}/2}^{T_{ps}/2} |G_r\left(\tau - \frac{2R_0}{c_0}\right)|^2 d\tau}, \quad k \neq j \tag{4.62}$$

where τ_d is the time delay between two adjoining subpulses, T_{ps} is the subpulse duration, and $G_r(\tau)$ is the receive antenna's elevation pattern.

4.4.2.3 NESZ Performance Analysis

The target reflected power available at the near-space vehicle-borne receiver antenna is determined by

$$P_r = \frac{P_t \cdot G_t(\eta_i)}{4 \cdot \pi \cdot R_0^2(\eta_i)} \cdot \frac{\sigma_0}{4 \cdot \pi \cdot R_0^2(\eta_i)} \cdot \frac{\lambda^2 \cdot G_r(\eta_i)}{4 \cdot \pi} \tag{4.63}$$

where P_t is the transmit peak power, $G_t(\eta_i)$ and $G_r(\eta_i)$ with η_i incidence angle are the transmit and receive antenna gain respectively, and $R_0(\eta_i)$ is the slant range. As the total data samples are processed coherently to produce a single imaging resolution cell, the receiver thermal noise samples can be taken as independent from sample to sample within each pulse, and from pulse to pulse. After coherent range and azimuth compression, the final image SNR can be represented by

$$SNR_{image} = \frac{P_t \cdot G_t(\eta_i) \cdot G_r(\eta_i) \cdot \lambda^3 \cdot c_0 \cdot T_p \cdot PRF \cdot \sigma_0}{256 \cdot \pi^3 \cdot R_0^3(\eta_i) \cdot v_s \cdot \sin(\eta_i) \cdot K \cdot T_{sys} \cdot B_n \cdot F_n \cdot L_f} \tag{4.64}$$

The NESZ can then be calculated by

$$NESZ = \frac{256 \cdot \pi^3 \cdot R_0^3(\eta_i) \cdot v_s \cdot \sin(\eta_i) \cdot K_0 \cdot T_{sys} \cdot B_n \cdot F_n \cdot L_s}{P_t \cdot G_t(\eta_i) \cdot G_r(\eta_i) \cdot \lambda^3 \cdot c_0 \cdot T_p \cdot PRF}. \tag{4.65}$$

4.4.3 Conceptual Examples and Simulation Results

To evaluate the quantitative performance of near-space vehicle-borne SAR with reflector antenna for HRWS remote sensing, we consider an example system. The SAR operates in X-band with a carrier frequency of 10 GHz. Table 4.2 gives the corresponding system parameters. Note that, a far-field, flat-earth, free-space, and single polarization model is also assumed. It is also assumed that the SAR moves at a constant velocity and operates in stripmap mode. On transmit activating all elements gives a wide low gain beam illuminating the complete swath. On receive the energy returned from a narrow portion of the ground illuminates the entire reflector, but is focused on individual elements of the feed aperture because the SCORE technique is employed in the elevation.

It is worthwhile to compare the RASR performance of the reflector antenna and SCORE technique combined SAR to conventional single beam SAR. Using the system parameters listed in Table 4.2, Fig. 4.11 shows the comparative RASR performance as a function of slant range. Note that equal parameters of $PRF (= 4,000\,\text{Hz})$

Table 4.2 Near-space vehicle-borne reflector antenna SAR system parameters

Parameters	Values	Units
Carrier frequency	10	GHz
Platform velocity	500	m/s
Platform altitude	20	km
Reflector antenna length	1.5	m
Reflector antenna width	0.8	m
Earth radius	6370	km
Minimum slant range	40	km
Maximum slant range	100	km
Transmit peak power	1000	W
Pulse duration time	10	μs
Pulse bandwidth	100	MHz
Range sampling frequency	150	MHz
Number of azimuth channel	3	–
Number of subswath	4	–

and subswath width are assumed in the simulations. For efficient SAR imaging, the RASR should be smaller than -20 dB. For example, the calculated RASR are -35 dB for the reflector antenna and SCORE technique combined case and -12 dB for general single-beam case, respectively. These results clearly show that a significant RASR performance improvement is obtained for the reflector antenna and SCORE technique combined approach. This means that a wider swath is possible for the near-space vehicle-borne SAR with reflector antenna, however, without decreasing the operating *PRF* which means that a higher azimuth resolution can be obtained.

It is also worthwhile to compare the AASR performance of the reflector antenna SAR with the conventional single azimuth beam SAR. Consider the system parameters listed in Table 4.2, Fig. 4.12 gives the corresponding comparative AASR performance as a function of *PRF*. Note that $\Gamma_k = -10$ dB is assumed in the simulation examples. In SAR remote sensing applications, AASR is typically specified to be on the order of -20 dB, but a lower AASR is desired. It can be noted from Fig. 4.12 that the AASR is typically below -20 dB with a low operating PRF requirement. This means that a wider swath can be obtained.

Another imaging performance is the NESZ, which is given in Fig. 4.13. Note that $L_s = 3$ dB and $F_n = 3$ dB are assumed in the simulation. The notches in the curve shape are caused by the switching of the active elements and the corresponding pattern switching. We can see that, for the system parameters given in Table 4.2 the swath width ranges from -54 dB to -44 dB. When compared to the reflector antenna-based spaceborne SAR investigated in [48], here the *NESZ* performance is improved by 20 dB. The reason is that the near-space vehicles are 10–20 times closer to the targets than LEO satellites.

Unlike conventional planar antenna SARs in which complex multiplication and summation are required to form a time varying beam, digital threshold detectors are employed in the reflector antenna SAR because the reflectivity variation in different scenes causes a variation of the average power level at the receiver [48]. The threshold

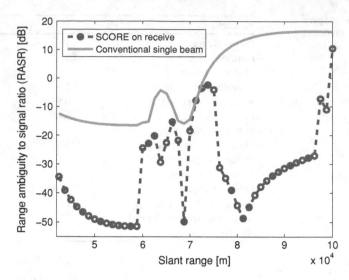

Fig. 4.11 Comparative RASR results between SCORE on receive and conventional single beam as a function of slant range

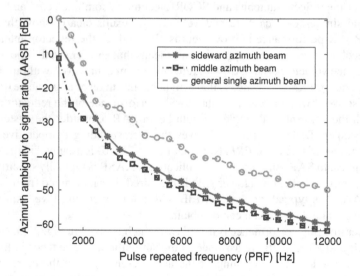

Fig. 4.12 Comparative AASR results conventional single azimuth beam SAR as a function of PRF

should ensure that at each time instance only the information relevant channels are summed up. However, in this case the azimuth Doppler spectrum will be undulated by the azimuth antenna pattern, as shown in Fig. 4.14. Consequently, the imaging performance will be degraded. This problem can be resolved by inverse filtering the azimuth multichannel signals.

Fig. 4.13 NESZ results as a function of slant range

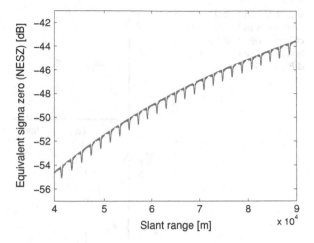

Fig. 4.14 Impact of azimuth antenna pattern on reflector antenna SAR Doppler spectra

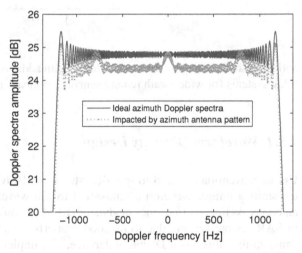

4.5 Near-Space HRWS Remote Sensing via Waveform Diversity

Recently, researches on multiple-input and multiple-output (MIMO) SAR have been drawing more and more attention [53–56]. MIMO is a technique used previously in communications to increase data throughout and link range without additional bandwidth or transmit power. Given that MIMO SAR is in its infancy, there is no clear definition of what it is. It is generally assumed that independent signals are transmitted through different antennas, and that these signals, after propagating through the environment, are received by multiple antennas. Generally speaking, MIMO SAR has two advantages when compared to the traditional SARs. The first advantage lies in spatial diversity gain and flexible spatial transmit beampattern design. The second advantage is resolution improvement. These are the reasons why MIMO SAR system

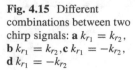

Fig. 4.15 Different combinations between two chirp signals: **a** $k_{r_1} = k_{r_2}$, **b** $k_{r_1} = k_{r_2}$, **c** $k_{r_1} = -k_{r_2}$, **d** $k_{r_1} = -k_{r_2}$

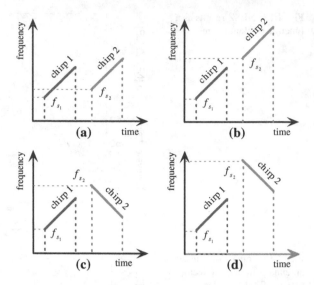

and the associated exploitation methodology can address the shortcomings of current SAR systems for wide-swath remote sensing applications.

4.5.1 Waveform Diversity Design

Unlike conventional phased array SAR systems, in MIMO SAR systems each antenna transmits a unique waveform, orthogonal to the waveforms transmitted by other antennas. As linearly frequency modulated (LFM) waveform has been widely utilized in SAR remote sensing due to its good properties such as high range resolution, constant modulus, good Doppler tolerance, and implementation simplicity, from a practical point of view we think that MIMO SAR should use LFM-based waveforms, so as to reduce the tough requirement of high transmit power. An adaptive LFM waveform diversity was proposed in [57], only the simple up- and down-chirp signals are allowed. It can be extended into orthogonal frequency diversion multiplexing (OFDM) chirp waveform.

From a practical point of view, suppose the chirp signals have equal frequency bandwidth and inverse or equal chirp rate (i.e., $|k_{r1}| = |k_{r2}|$), there are four different combinations between two chirp signals, as shown in Fig. 4.16. For MIMO SAR range compression, the auto-correlation between identical chirp signals is used to evaluate the range resolution, and the cross-correlation between different chirp signals is crucial for suppressing the range ambiguity.

As an example, Fig. 4.16 shows the correlation characteristics of the different chirp combinations shown in Fig. 4.15. Note that the parameters of bandwidth $B = 10\,\text{MHz}$, starting frequency $f_{s2} = 0\,\text{Hz}$ and pulse duration $T_p = 5 \cdot 10^{-6}\text{s}$ are

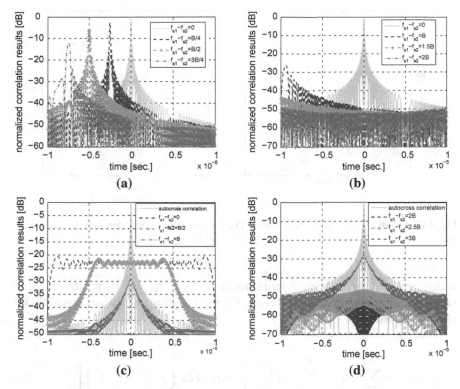

Fig. 4.16 Correlation characteristics of the different chirp combinations: **a** and **b** have equal chirp rates, **c** and **d** have inverse chirp rate

assumed in the simulations. For MIMO SAR imaging, a quantity of importance is the relative level between the correlation of identical chirp signals and the correlation of different chirp signals. Hence, the chirp waveforms with high cross-correlational suppression are desired. The correlation results for different starting frequencies and equal chirp rate are shown in Figs. 4.16a and 4.16b. It is obvious that the performance of cross-correlation suppression improves with the increase the separation between two starting frequencies (i.e. $f_{si} - f_{sj}$). The correlation results for different starting frequencies and inverse chirp rate are shown in Figs. 4.16c and 4.16d. It is also obvious that the maximum and occupied time of the correlation will decrease with the increase of the separation between two starting frequencies.

It is deduced from Fig. 4.16 that using chirp signals with equal chirp rate and $|f_{si} - f_{sj}| \geq 2B$ offers a satisfactory suppression of the cross-correlation components. However, for a specific requirement of frequency bandwidth, this also means a wider total transmit/receive bandwidth for the MIMO SAR radio frequency hardware systems. Fortunately, the chirp signals with inverse chirp rate and adjacent starting frequency (see Fig. 4.17d $f_{si} - f_{sj} = 2B$) can provide a good cross-correlational suppression. Therefore, it is possible to use an arbitrary number of chirp signals

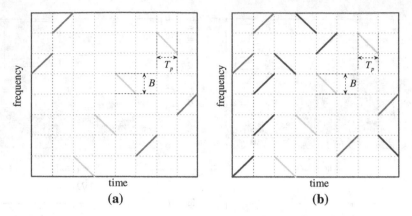

Fig. 4.17 Example OFDM chirp waveforms. The number of the chirp signals is for illustration only

which occupy adjacent starting frequency (i.e., $f_{si} - f_{sj} = 2B$) for inverse chirp rate or non-overlapping frequency bands. Figure 4.17 illustrates two example OFDM chirp waveforms.

They can be expressed as

$$s_k(t) = \sum_{k=1}^{K} s_k(t) = \sum_{k=1}^{K} \text{rect}\left[\frac{t}{T_p}\right] \cdot \exp\left\{j2\pi \left(f_{s_k} t + \frac{1}{2} k_{r_k} t^2\right)\right\} \qquad (4.66)$$

where f_{s_k} and k_{r_k} denote the starting frequency and chirp rate for the k chirp signal, respectively. We use the ambiguity function of the OFDM chirp waveforms defined in [58]

$$\chi(t_d, f_d) \triangleq \sum_{k=0}^{K} \sum_{k'=0}^{K} \int_{-\infty}^{\infty} s_k(t) s_{k'}^*(t + t_d) e^{j2\pi f_d t} dt \qquad (4.67)$$

where f_d denotes the Doppler frequency. This equation can be calculated by numerical integral.

Consider the OFDM chirp waveforms shown in Fig. 4.17 and suppose the following parameters: $B = 10\,\text{MHz}$, $T_p = 10\mu\text{s}$, $f_{s_1} = 50\,\text{MHz}$, $f_{s_2} = 70\,\text{MHz}$, $f_{s_3} = 10\,\text{MHz}$, $f_{s_4} = 30\,\text{MHz}$, $f_{s_5} = 50\,\text{MHz}$, $f_{s_6} = 10\,\text{MHz}$, $f_{s_7} = 70\,\text{MHz}$, $f_{s_8} = 30\,\text{MHz}$, Fig. 4.18 shows the ambiguity function characteristics. As the value of $\chi(0, 0)$ represents the matched filtering output without any mismatch, the sharper the function $|\chi(\tau, f_d)|$, the better the range and azimuth (Doppler) resolution that can be obtained for the radar system. The results show that the OFDM chirp waveform has a satisfactory ambiguity function performance in range resolution and Doppler frequency resolution. This is particularly valuable for the subsequent MIMO SAR image formation processing.

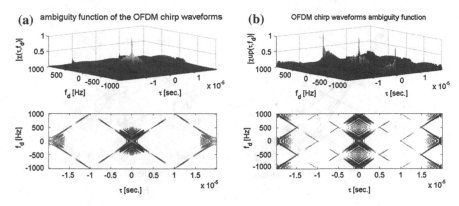

Fig. 4.18 Ambiguity functions of the OFDM chirp waveforms shown in Fig. 4.17

4.5.2 MIMO SAR-Based Wide-Swath Remote Sensing

The multiple antennas in the MIMO SAR systems can be arranged in along-track (azimuth, flight direction), cross-track (elevation, perpendicular to along-track), or in both dimensions. In the following discussions the multiple antennas are placed only in the cross-track direction. If each antenna transmits a sufficiently wideband signal, each transmit antenna and receive antenna pair can then act as an individual SAR, and the MIMO SAR can then serve as multiple SARs operating independently. The cumulative wide bandwidth can offer an extremely high range resolution. This architecture has the potential to increase azimuth resolution while enabling high gain on target.

4.5.2.1 System Scheme

As shown in Fig. 4.19a, the basic idea is to form multiple transmit and receive beams which are steered toward different subswaths. This is different from a phased array radar because orthogonal transmit signals are employed in the MIMO SAR system. Multiple transmit and receive beams can then be formed by grouping the array elements into multiple groups, each forming a transmit or receive subaperture. These subapertures can be disjointed or overlapped in space, but disjoint subapertures are assumed. Suppose the transmit array has M collocated elements, whereas the receive array has N collocated elements. A total of $N_t < M$ transmit beams can then be formed. The waveforms used in any two subapertures are orthogonal, but the same waveform is used in each subaperture. As shown in Fig. 4.19b (shows only two beams, it is for illustration only), N_t transmit subapertures are used to form N_t different directional beams, each modulated by a different and orthogonal function $s_k(t)$, $k = 1, 2, \ldots, N_t$. As the signal transmitted by the kth subaperture steering toward an angle θ_0 is given by $b_k^*(\theta_0)s_k(t)$ with $b_k(\cdot)$ the steering vector,

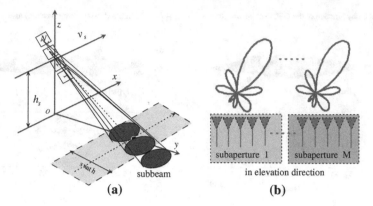

Fig. 4.19 Geometry of near-space vehicle-borne MIMO SAR with multi-antenna in elevation

the equivalent baseband signal seen at a location with angle θ_i can be expressed as $b_k^T(\theta_i)b_k^*(\theta_0)s_k(t)$, where $(\cdot)^T$ denotes a transpose operator. Correspondingly, the overall signal is a superposition of the transmissions from all N_t subapertures

$$s_{N_s}(t) \triangleq \sum_{k=1}^{N_t} b_k^T(\theta_i)b_k^*(\theta_0)s_k(t) \tag{4.68}$$

In this way, multiple virtual transmit beams can be formed.

As the transmitted waveforms are orthogonal, the receivers can form N_r separate receive patterns in elevation, each with nulls in the directions of $N_r - 1$ ambiguities, and a beam in the direction of the appropriate echo. Like the multiaperture SAR discussed previously, this means that we can divide the whole swath into N_r subswaths

$$\frac{c_0}{2}\left(\frac{i+k-1}{PRF} + T_p + \Delta T_{tr}\right) < R_k < \frac{c_0}{2}\left(\frac{i+k}{PRF} - T_p\right), \leq k \leq N_r \tag{4.69}$$

For M transmit array elements and N receive array elements, there will be $M \times N$ different echoes, each is weighted by a different amplitude and phase. Each receive channel signal is separately demodulated, digitized, and stored. Next, a posteriori, digital beamforming on receive can then be carried out by a joint spatiotemporal processing. In this way, multiple pairs of virtual transmit and receive beams can be formed simultaneously in the direction of a wanted subswath.

4.5.2.2 Matched Filtering and Multibeam Forming

Multibeam forming in elevation must be carried out for subsequent image formation processing. The beamforming filter coefficients should be optimally designed, so

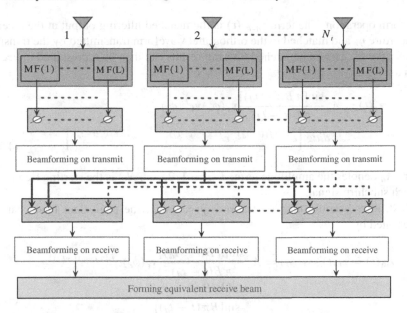

Fig. 4.20 The scheme of MIMO SAR multi-beam forming in elevation

that the desired beam direction to the intended subswath can be obtained. To reach this aim, the multi-beam forming scheme shown in Fig. 4.20 is used.

Suppose the transmit array is a uniform linear array, the transmit signals expressed in Eq. 4.68 can be reformed into

$$\mathbf{s}(t) = \mathbf{v_t}^T(\phi_t) \left[s_1(t), s_2(t), \ldots, s_{N_t}(t) \right]^T \tag{4.70}$$

where $\mathbf{v_t}(\phi_t) = [1, e^{-j2\pi[(d \sin \phi_t)/\lambda]}, \ldots, e^{-j2\pi(N_t-1)[(d \sin \phi_t)/\lambda]}]^T$ with ϕ_t the incidence angle is the transmit array response vector, N_t is the number of the transmit subapertures, $[s_1(t), s_2(t), \ldots, s_{N_t}(t)]$ is the signals transmitted by the transmitter. The receive signals can be written as a vector $\mathbf{x}(t) = [x_1(t), x_2(t), \ldots, x_{N_r}(t)]^T$ (N_r is the number of the receive subapertures)

$$\mathbf{x}(t) = \mathbf{v_r}(\phi_r)\mathbf{v_t}^T(\phi_t)\mathbf{s}(t - \tau_d) \tag{4.71}$$

where $\mathbf{v_r}(\phi_r) = [1, e^{-j2\pi[(d \sin \phi_r)/\lambda]}, \ldots, e^{-j2\pi(N_r-1)[(d \sin \phi_r)/\lambda]}]^T$ and t_d represent respectively the receive array response vector and the time it takes the signal to travel the transmitter-target-receiver distance.

The matched filtering of each receive channel can be represented by

$$z_{n_r n_t}(t) = x_{n_r}(t) \otimes s_{n_t}^*(t) = F^{-1}\left\{ F\{x_{n_r}(t)\} \cdot F\{s_{n_t}^*(t)\} \right\} \tag{4.72}$$

where $n_t \in [1, 2, \ldots, N_t]$, $n_r \in [1, 2, \ldots, N_r]$, the $\otimes$, F^{-1} and F denote respectively the convolution operation, inverse Fourier transform operation, and Fourier transform operation. The term $z_{n_r n_t}(t)$ is the matched filtering output at the receive subaperture n_r and matched to the orthogonal waveform transmitted by the transmit subaperture n_t. The total matched filtering results can then be represented in vector form as

$$
\begin{aligned}
\mathbf{z_r}(t) = A_a \frac{\sin[\pi B(t - t_d)]}{\pi B(t - t_d)} \mathbf{v_r}(\phi_r) \mathbf{v_t}^T(\phi_t) \\
\times diag \left\{ e^{j2\pi f_1(t - t_d)}, e^{j2\pi f_2(t - t_d)}, \ldots, e^{j2\pi f_{N_t}(t - t_d)} \right\}
\end{aligned}
\tag{4.73}
$$

where A_a denotes the amplitude term, $f_i (i = 1, 2, \ldots, N_t)$ is the starting frequency of each subchirp signal.

Next, after beamforming on transmit processing, the equivalent beam can be represented by

$$
\begin{aligned}
\mathbf{z_{in}}(t) = \mathbf{z_r}(t) \cdot \mathbf{v_t}(\phi_t) = A_a \frac{\sin[\pi B(t - t_d)]}{\pi B(t - t_d)} \mathbf{v_r}(\phi_r) e^{j2\pi f_{\min}(t - t_d)} \\
\times \frac{\sin[N_t B\pi(t - t_d)]}{\sin[B\pi(t - t_d)]}
\end{aligned}
\tag{4.74}
$$

where $\mathbf{v_t}(\phi_t)$ is the transmit steering vector and $f_{\min}$ is the smallest frequency center among the N_t orthogonal subchirp signals. Correspondingly, for the nth receive subaperture we can get

$$
\begin{aligned}
\mathbf{z_{out,n}}(t) &= \mathbf{v_{r,n}^T}(\phi_r) \cdot \mathbf{z_{in}}(t) \\
&= N_r A_a \frac{\sin[\pi B(t - t_d)]}{\pi B(t - t_d)} \cdot \frac{\sin[N_t B\pi(t - t_d)]}{\sin[B\pi(t - t_d)]} \cdot e^{j2\pi f_{\min}(t - t_d)}
\end{aligned}
\tag{4.75}
$$

where $\mathbf{v_{r,n}}(\phi_r)$ is the steering vector for the nth receive subaperture. Then there is

$$
|\mathbf{z_{out,n}}(t)| = \left| N_r A_a \frac{\sin[\pi B(t - t_d)]}{\pi B(t - t_d)} \cdot \frac{\sin[N_t B\pi(t - t_d)]}{\sin[B\pi(t - t_d)]} \right|.
\tag{4.76}
$$

Therefore, if the matched filtering outputs are fed into multiple digital filters with their coefficients designed for different beam-pointing directions, multiple pairs of virtual transmit and receive beams in predetermined directions can be formed simultaneously. Moreover, high range resolution is also obtained. We conclude also that a smaller ($1/L_r$ for one subaperture and $1/N_r$ for the whole aperture) RCS target can be detected for the MIMO SAR. Correspondingly, for a given requirement of SAR image SNR, a lower peak transmit power or average transmit power is required.

4.5.2.3 Numerical Simulation Results

According to multi-transmission and multi-reception in the elevation scheme described previously, eight subchirp signals are transmitted from the eight transmit

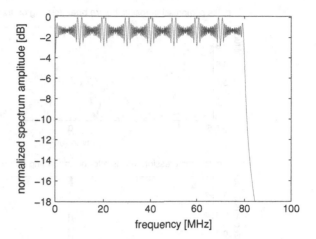

Fig. 4.21 The returned signal spectrum received by a subaperture

subarrays simultaneously and received by four subapertures. The reflected signals will be received by all the receive subantennas.

For simplicity, suppose there is a point target at the swath center and with an azimuth distance of 0 m, Fig. 4.21 shows the signal spectrum received by a single subantenna. It can be noted that, after beamforming on transmit, an echo with wide frequency width can then be received by each receive subantenna. As the transmit sub-chirp signals are orthogonal, their matched filtering can be implemented separately with the corresponding reference functions. In this way, the final range resolution can be approximately improved by a factor of N_t, i.e., the number of the subchirp signals transmitted simultaneously, as shown in Fig. 4.17. As an example, suppose the inter-element spacing is half wavelength, Fig. 4.22 shows the magnitude of the directional gain for an example subaperture with $L_t = 8$ elements. Next, one or multiple beams can be obtained by cooperative digital beamforming on receive. Thereafter, the subswath data can be coherently combined into the raw data for unambiguous wide-swath imaging.

4.5.3 Space-Time Coding MIMO SAR for High-Resolution Imaging

In this section, we further proposed a space-time coding MIMO SAR for high-resolution remote sensing. As shown in Fig. 4.23, the approach employs MIMO configuration in elevation direction and a space-time coding scheme in azimuth direction, along with OFDM waveform diversity and DPCA techniques. The basic idea is to divide the total transmit antenna elements into multiple groups, each forming a subbeam, thus offering several benefits, including improved ambiguity suppression for wide-swath imaging, improved SNR, and flexible operational configuration.

pulse compression results before beamforming on transmit subaperture

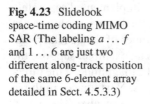

pulse compression results after beamforming on transmit subaperture

Fig. 4.22 The final matched filtering results for one subaperture: the above one is before beamforming on transmit subaperture, the below one is after beamforming on transmit subaperture

Fig. 4.23 Slidelook space-time coding MIMO SAR (The labeling $a \ldots f$ and $1 \ldots 6$ are just two different along-track position of the same 6-element array detailed in Sect. 4.5.3.3)

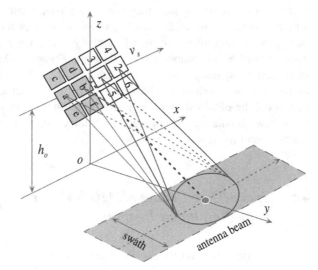

4.5.3.1 Space-Time Coding Transmission in the Azimuth Direction

The simplest Alamouti code is suitable for the space-time coding MIMO-OFDM SAR system with two transmit antennas [59], which is based on the following 2×2 orthogonal matrix

$$\mathbf{S}(s_1, s_2) = \begin{bmatrix} s_1 & s_2^* \\ s_2 & -s_1^* \end{bmatrix} \tag{4.77}$$

where $(\cdot)^*$ denotes the complex conjugate operator, s_1 and s_2 denote the two transmitted signals. The entry at the ith row and jth column represents the signal to be transmitted from the ith antenna at the jth time interval. That is to say, two different signals s_1 and s_2 are transmitted simultaneously from the antennas 1 and 2 respectively during the first signal period, following which the signals s_2^* and $-s_1^*$ are transmitted from antennas 1 and 2 respectively. We assume that the target's RCS remains constant over consecutive signal periods. The channel response for the nth receive antenna can then be represented by

$$\mathbf{H} = \begin{bmatrix} h_{1,n}, h_{2,n} \end{bmatrix} \tag{4.78}$$

where $h_{1,n}$ and $h_{2,n}$ denote the channel response of the two transmitted signals, respectively.

The signals received by the nth receive antenna over consecutive signal periods are

$$\begin{cases} r_{1n} = s_1 \otimes h_{1,n} + s_2 \otimes h_{2,n} + n_{1,n} \\ r_{2n} = s_2^* \otimes h_{1,n} - s_1^* \otimes h_{2,n} + n_{2,n} \end{cases} \tag{4.79}$$

where $n_{1,n}$ and $n_{2,n}$ denote individual additive white Gaussian noise. This signal model is formulated in time domain. Transforming them into frequency domain yields

$$\begin{bmatrix} R_{1n} \\ R_{2n} \end{bmatrix} = \begin{bmatrix} S_1 & S_2 \\ S_2^* & -S_1^* \end{bmatrix} \cdot \begin{bmatrix} H_{1,n} \\ H_{2,n} \end{bmatrix} + \begin{bmatrix} N_{1,n} \\ N_{2,n} \end{bmatrix} \tag{4.80}$$

where R_{1n} (R_{2n}), S_1 (S_2), S_2^* (S_1^*), $H_{1,n}$ ($H_{2,n}$) and $N_{1,n}$ ($N_{2,n}$) denote the Fourier transforming representation of r_{1n} (r_{2n}), s_1 (s_2), s_2^* (s_1^*), $h_{1,n}$ ($h_{2,n}$) and $n_{1,n}$ ($n_{2,n}$), respectively. Next, the received radar echoes should be separated by decoding processing. As the transmitted signal matrix is known for both the transmitter and the receiver, the decoding matrix can be easily constructed as

$$\mathbf{D} = \begin{bmatrix} S_1^* & S_2 \\ S_2^* & -S_1 \end{bmatrix} \tag{4.81}$$

Since S_1 and S_2 are orthogonal, the decoded signals can be represented by

$$\begin{bmatrix} R_{1n}' \\ R_{2n}' \end{bmatrix} = \begin{bmatrix} S_1^* & S_2 \\ S_2^* & -S_1 \end{bmatrix} \cdot \begin{bmatrix} R_{1n} \\ R_{2n} \end{bmatrix} = \begin{bmatrix} (|S_1|^2 + |S_2|^2)H_{1,n} \\ (|S_1|^2 + |S_2|^2)H_{2,n} \end{bmatrix} + \begin{bmatrix} N_{1,n} - N_{2,n} \\ N_{1,n} + N_{2,n} \end{bmatrix}. \tag{4.82}$$

It is noted from Eq. 4.82 that, after Alamouti decoding, the two received signals have the same transmit information ($|S_1|^2 + |S_2|^2$) and dependent channel response. It can also be concluded that the Alamouti scheme extracts a diversity order of 2 if two orthogonal signals are transmitted with the same power means $|S_1|^2 = |S_2|^2$. After matched filtering, the reconstructed R_{1n}' and R_{2n}' achieve a gain of 12 dB while the noise level w 6 dB. Thus, the equivalent diversity gain is 6 dB.

4.5.3.2 MIMO Configuration in the Elevation Direction

In the elevation direction, the total antenna array elements are divided into multiple subarrays. These subarrays can be disjoint or overlapping in space, but here disjoint subarrays are assumed. Orthogonal waveforms are transmitted from different subarrays to illuminate a wide swath. Each receive subarray can receive all the reflected signals. If N_t transmit subarrays and N_r receive subarrays are employed, there will be $N_t \times N_r$ different returns for the receiver. For the mth ($m = 1, 2, \ldots, N_t$) transmit subarray and nth ($n = 1, 2, \ldots, N_r$) receive subarray, the radar response or scattering function can be approximated as a realization of one random process denoted by $a_{n,m}$ ($m = 1, 2, \ldots, N_t; n = 1, 2, \ldots, N_r$). The random process is such that over the time period of the transmit signal duration, the measured amplitude should be the same, i.e. $|a_{n,m}|$ is infact a constant (for one set of simultaneous measurements employing different codes).

The received signal at the nth receive subarray due to the mth transmit waveform can be represented by

$$r_{n,m}(t) = s_m(t, n)a_{n,m} + n_{n,m}(t) \tag{4.83}$$

The term $s_m(t, n)$ denotes the signals transmitted by the mth subarray and received by the nth subarray. The term $n_{n,m}(t)$ is an additive noise process being independent of the response function $a_{n,m}$. As there are N_t transmit signals, the received signals at the nth receive subarray are the linear combinations of all such signals

$$\mathbf{r_n} = \sum_{m=1}^{N_t} r_{n,m}(t) = \sum_{m=1}^{N_t} s_m(t, n)a_{n,m} + \sum_{m=1}^{N_t} n_{n,m}(t) = \mathbf{s}(n)\mathbf{a}_n + \mathbf{n}_n \tag{4.84}$$

The terms $\mathbf{s}(n)$, $\mathbf{a}_n$ and $\mathbf{n}_n$ are the matrices whose columns are the $s_m(t, n)$, $a_{n,m}$ and $n_{n,m}(t)$, respectively. As there are a total of N_r receive subarrays, the data from each subarray can be composed into a single vector

$$\mathbf{R} = {\mathbf{r}_1^T, \mathbf{r}_2^T, \ldots, \mathbf{r}_{N_r}^T}^T = \mathbf{S} \cdot \mathbf{A} + \mathbf{N} \tag{4.85}$$

where T denotes the transpose, $\mathbf{S} = diag[\mathbf{s}(1), \mathbf{s}(2), \ldots, \mathbf{s}(N_r)]$, $\mathbf{A} = [\mathbf{a}_1, \mathbf{a}_2, \ldots, \mathbf{a}_{N_r}]^T$ and $\mathbf{N} = [\mathbf{n}_1, \mathbf{n}_2, \ldots, \mathbf{n}_{N_r}]$. Equation 4.85 is just the signal model used to describe the received space-time coding MIMO SAR data in the elevation direction.

4.5.3.3 Relations and Models for Azimuth Signal Processing

In the azimuth direction, the space-time coding scheme results in equivalent phase centers when the radar moves from one position to another. This enables a coherent

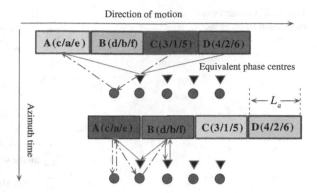

Fig. 4.24 Extended DCPA scheme for the space-time coding MIMO SAR system

combination of the subsampled signals with the DPCA technique, which synthesizes multiple receive beams that are displaced in the along-track direction. It implies that we can broaden the azimuth beam from the diffraction-limited width, giving rise to an improved resolution without having to increase the system operating PRF.

Figure 4.24 shows the extended DCPA scheme for the space–time coding MIMO SAR system shown in Fig. 4.23, in which 6 transmit subarrays and 12 receive subarrays are employed. With the space-time coding scheme described previously, at azimuth time η_1 the orthogonal signals $\mathbf{X_A}(t) = [x_{a1}(t), x_{a2}(t), x_{a3}(t)]^T$ and $\mathbf{X_B}(t) = [x_{b1}(t), x_{b2}(t), x_{b3}(t)]^T$ are transmitted from the subarrays (1, 3, 5) and (2, 4, 6), respectively. Similarly, at azimuth time η_2 the orthogonal signals $\mathbf{X_B^*}(t) = [x_{b1}^*(t), x_{b2}^*(t), x_{b3}^*(t)]^T$ and $-\mathbf{X_A^*}(t) = [-x_{a1}^*(t), -x_{a2}^*(t), -x_{a3}^*(t)]^T$ are transmitted from the subarrays (a, c, e) and (b, d, f), respectively. After beamforming processing in the elevation direction, it is possible to define the " two-way" phase centers as the mid-points between the equivalent transmit beam phase center and the equivalent receive beam phase center. All goes well as the radar samples are collected by the equivalent transmit and receive beams with phase centers collocated in the two-way phase centers. We aim to collect the radar returns at the time when their equivalent two-way phase centers occupy the at same spatial position, along with the platform trajectory.

When the DCPA condition is matched, after beamforming processing in elevation we consider the geometry of equivalent beam phase centers, as shown in Fig. 4.25. The term $\alpha(\tau)$ is the equivalent beam instantaneous squint angle at azimuth time τ, and R_c is the closest range to a given point target when the MIMO SAR platform moves along its trajectory. For the signals transmitted at azimuth time τ_1, the equivalent beam two-way phase centers between the equivalent transmit beam D and the equivalent receive beam A can be represented by

$$R_{DA}(\tau_1) = \sqrt{\left[\frac{3L_a}{2} - R_c\tan(\alpha(\tau_1))\right]^2 + R_c^2} + \sqrt{\left[\frac{3L_a}{2} + R_c\tan(\alpha(\tau_1))\right]^2 + R_c^2}$$

(4.86)

Fig. 4.25 Geometry of equivalent beam phase centers

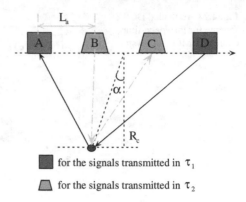

for the signals transmitted in τ_1

for the signals transmitted in τ_2

where L_a is the distance between subarrays in the azimuth direction. Similarly, for signals transmitted by the equivalent beam B at azimuth time η_2 and received by the equivalent beam C we have

$$R_{BC}(\tau_2) = \sqrt{\left[\frac{L_a}{2} - R_c \tan(\alpha(\tau_2))\right]^2 + R_c^2} + \sqrt{\left[\frac{L_a}{2} + R_c \tan(\alpha(\tau_2))\right]^2 + R_c^2}$$
(4.87)

The corresponding equivalent phase difference between $R_{DA}(\tau_1)$ and $R_{BC}(\tau_2)$ is

$$\Delta\Phi(\tau_1, \tau_2) = \frac{4\pi}{\lambda}[R_{DA}(\tau_1) - R_{BC}(\tau_2)]$$
(4.88)

As the equivalent receive beam is assumed to be coincident in the far-field region and is of the same width as the equivalent transmit beam, for two successive pulse repeated intervals (PRI = 1/PRF).

$$\alpha(\tau_1) \approx \alpha(\tau_2)$$
(4.89)

As an example, assuming an X-band MIMO SAR system because this carrier frequency is popular in current spaceborne SAR systems with the following parameters: $\lambda = 0.03$ m, $R_c = 700$ km and $L_a = 4$ m, Fig. 4.26 shows the corresponding equivalent phase difference $\Delta\Phi(\tau_1, \tau_2)$ as a function of the instantaneous squint angle α. We note that the phase difference is small and may be neglected during subsequent image formation processing. More importantly, in this case the channel responses $H_{1,n}$ and $H_{2,n}$ at azimuth time τ_1 and τ_2 can be seen as being equivalent. Equation 4.82 can then be simplified into

$$\begin{bmatrix} R_{1n}' \\ R_{2n}' \end{bmatrix} = \begin{bmatrix} (|S_1|^2 + |S_2|^2)H_{1,n} \\ (|S_1|^2 + |S_2|^2)H_{1,n} \end{bmatrix} + \begin{bmatrix} N_{1,n} - N_{2,n} \\ N_{1,n} + N_{2,n} \end{bmatrix}$$
(4.90)

Therefore, the signals received by the subantenna n can be recombined, so as to further improve its SNR performance.

Fig. 4.26 Equivalent phase difference $\Delta\Phi(\tau_1, \tau_2)$ between two successive PRI transmission as a function of the instantaneous squint angle

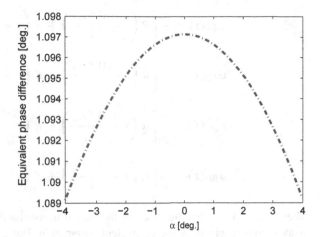

The range from the equivalent MIMO SAR phase center to an arbitrary point target can then be represented by

$$R(\tau) = \sqrt{R_0^2 + v_s^2(\tau - \tau_0)^2 - 2v_s(\tau - \tau_0)R_0\cos(\theta)} \qquad (4.91)$$

where R_0 is the slant-range to broadside of antenna center, η_0 is the time at target broadside, and θ is the angle of the target off broadside. In SAR applications, cubic and higher order terms may be ignored in the azimuth phase history of targets. A general criterion for this condition to hold is that this term causes less than $\pi/4$ excursion the phase over the aperture synthesis time. In this case, Eq. 4.91 can be simplified into

$$R(\tau) = R_0 - v_s\cos(\theta)(\tau - \tau_0) + \frac{v_s^2\sin^2(\theta)(\tau - \tau_0)^2}{2R_0} \qquad (4.92)$$

Therefore, for the case illustrated at the bottom of Fig. 4.24 the Doppler phase histories of the four received subarrays can be represented, respectively, by

$$\psi_{AA}(\tau) = \frac{1}{\lambda}\left[R\left(\tau - \frac{1.5L_a}{v_s}\right) + R\left(\tau - \frac{1.5L_a}{v_s}\right)\right] \qquad (4.93a)$$

$$\psi_{AB}(\tau) = \frac{1}{\lambda}\left[R\left(\tau - \frac{1.5L_a}{v_s}\right) + R\left(\tau - \frac{0.5L_a}{v_s}\right)\right] \qquad (4.93b)$$

$$\psi_{AC}(\tau) = \frac{1}{\lambda}\left[R\left(\tau - \frac{1.5L_a}{v_s}\right) + R\left(\tau + \frac{0.5L_a}{v_s}\right)\right] \qquad (4.93c)$$

$$\psi_{AD}(\tau) = \frac{1}{\lambda}\left[R\left(\tau - \frac{1.5L_a}{v_s}\right) + R\left(\tau + \frac{1.5L_a}{v_s}\right)\right] \qquad (4.93d)$$

$$\psi_{BA}(\tau) = \frac{1}{\lambda}\left[R\left(\tau - \frac{0.5L_a}{v_s}\right) + R\left(\tau - \frac{1.5L_a}{v_s}\right)\right] \tag{4.93e}$$

$$\psi_{BB}(\tau) = \frac{1}{\lambda}\left[R\left(\tau - \frac{0.5L_a}{v_s}\right) + R\left(\tau - \frac{0.5L_a}{v_s}\right)\right] \tag{4.93f}$$

$$\psi_{BC}(\tau) = \frac{1}{\lambda}\left[R\left(\tau - \frac{0.5L_a}{v_s}\right) + R\left(\tau + \frac{0.5L_a}{v_s}\right)\right] \tag{4.93g}$$

$$\psi_{BD}(\tau) = \frac{1}{\lambda}\left[R\left(\tau - \frac{0.5L_a}{v_s}\right) + R\left(\tau + \frac{1.5L_a}{v_s}\right)\right] \tag{4.93h}$$

where $\psi_{mn}(\tau)$ denotes the phase of the signal transmitted from the equivalent sub-array m and received by the equivalent subarray n. For example, Eq. 4.93c can be expanded into

$$\psi_{AC}(\tau) = \frac{1}{\lambda}\left[2R_0 - v_s\cos(\theta)\left(\tau - \frac{0.5L_a}{v_s} - \tau_0\right)\right.$$
$$\left. + \frac{v_s^2\sin^2(\theta)\left(\theta - \frac{0.5L_a}{v_s} - \theta_0\right)^2}{R_0}\right] + \frac{L_a^2\sin^2(\theta)}{R_0\lambda} \tag{4.94}$$

In near-space vehicle-borne SAR applications, L_a, R_0 and λ are typically on the orders of $1\,m$, $20\,km$ and $10\,cm$, respectively, hence the last term $L_a^2\sin^2(\theta)/R_0\lambda$ is small and can be ignored. We then have

$$\psi_{AC}(\tau) = \frac{2}{\lambda}R\left(\tau - \frac{0.5L_a}{v_s}\right) = \psi_{BB}(\tau) \tag{4.95}$$

In a similar manner, we can get

$$\psi_{AD}(\tau) = \psi_{BC}(\tau) = \frac{2}{\lambda}R(\tau) \tag{4.96}$$

and

$$\psi_{AA}(\tau) = \frac{2}{\lambda}R\left(\tau - \frac{1.5L_a}{v_s}\right) \tag{4.97a}$$

$$\psi_{BD}(\tau) = \frac{2}{\lambda}R\left(\tau + \frac{0.5L_a}{v_s}\right) \tag{4.97b}$$

$$\psi_{AB}(\tau) = \psi_{BA}(\tau) = \frac{2}{\lambda}R\left(\tau - \frac{1.0L_a}{v_s}\right) \tag{4.97c}$$

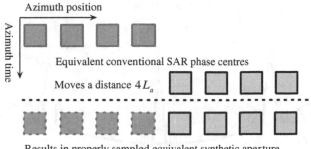

Fig. 4.27 The sampled equivalent synthetic aperture

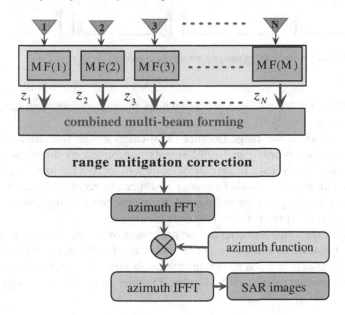

Fig. 4.28 MIMO SAR image formation processing via the rang-Doppler algorithm

To construct one synthetic aperture, the system must operate with a PRF which leads, after combination of data streams, to a properly sampled synthetic aperture appropriate to the beamwidth of the system. Once the DPCA condition is met, the operating PRF of the system is always equal to the Nyquist sampling rate for the diffraction-limited beamwidth of the antennas. According to Eqs. 4.95–4.97, Fig. 4.27 illustrates the corresponding equivalent phase centers. In this way, for a fixed PRF value, the antenna length of the MIMO SAR can remain roughly constant with increasing $N_{ea}/2$ (N_{ea} is the number of equivalent beams in azimuth, and 2 results from the Alamouti coding scheme) the achievable azimuth resolution. Alternatively, for a fixed azimuth resolution, an increase in swath coverage can be

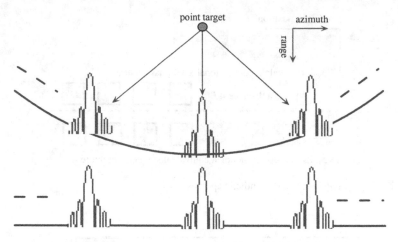

Fig. 4.29 Illustration of the range mitigation correction process

obtained. If the DCPA condition is not matched, the gathered azimuth samples will be spaced nonuniformly.

Figure 4.28 gives the range-Doppler (RD)-based image formation processing steps. As there are 6 orthogonal LFM signals transmitted by the transmit subantennas simultaneously, each receive subantenna will receive the total 6 signals. It is obvious that an echo with wide frequency width can be received by each subarray. As the transmitted signals are orthogonal, their matched filtering can be obtained separately with the corresponding reference functions.

Thereafter, the combined data can be processed by a uniform image formation algorithm. First, range mitigation correction is employed, as shown in Fig. 4.29. Then, the focused image can be obtained through azimuth compression processing which is like normal monostatic SAR range-Doppler image formation algorithms.

4.6 Conclusion

Near-space vehicles provide a potential to high-resolution and wide-swath SAR imaging which is a contradiction for conventional SARs. This chapter describes the potential configurations including multichannel in azimuth, multiaperture in elevation, and space-time coding MIMO SAR. The corresponding system schemes, mathematical relations, and signal models are formed. It is anticipated that near-space vehicle-borne multiantenna SARs, especially MIMO SAR, enable SAR systems to be operated with high flexibility and reconfigurability, thus ensuring previously unprecedented remote sensing performance. However, further research work is required. One future work is to develop high-precision image formation algorithms for MIMO SAR imaging. Another future work is synchronization processing. In this chapter, we considered only monostatic MIMO SAR. Bi- and multi-static MIMO SARs are also feasible, but this configuration will result in technical challenges that lie in the synchronization between the transmitters and receivers, which include spatial synchronization, time synchronization, and phase synchronization.

References

1. Li, Z.F., Wang, H.Y., Su, T., Bao, Z.: Generation of wide-swath and high-resolution SAR images from multichannel small spaceborne SAR systems. IEEE. Geosci. Remote. Sens. Lett. **2**, 82–86 (2005)
2. Currie, A., Brown, M.A.: Wide-swath SAR. IEE. Radar. Signal. Process. **139**, 122–135 (1992)
3. Curlander, J.C., McDonough, R.N.: Synthetic Aperture Radar: Systems and Signal Processing. John Wiley & Sons, Inc. (1991)
4. The paradigm shift of effects-based space: near-space as a combat space effects enabler. http://www.airpower.au.af.mi. Accessed May 2010
5. Suess M., Grafmuller B., Zahn R.: A novel high resolution, wide swath SAR system. In: Proceeings of IEEE Int Geosci Remote Sens Symposium, Sydney, Australia 1013–1015 (2001)
6. Suess M., Zubler M., Zahn R.: Performance investigation on the high resolution, wide swath SAR system. In: Proceedings of European Synthetic Aperture Radar Conference, Cologne, Germany 187–191 (2002)
7. Heer C., Soualle F., Zahn R., Reber R.: Investigations on a new high resolution wide swath SAR concept. In: Proceedings of IEEE Int Geosci Remote Sens Symp, Toulouse, France 521–523 (2003)
8. Gebert N., Krieger G., Moreira A.: High resolution wide swath SAR imaging— system performance and influence of perturbations. In: Proceedings of Int Radar Symp, Berlin, Germany 1–5 (2005)
9. Gebert, N., Krieger, G.: Azimuth phase center adaptation on transmit for high-resolution wide-swath SAR imaging. IEEE. Geosci. Remote. Sens. Lett. **6**, 782–786 (2009)
10. Gebert N., Krieger G., Younis M., Bordoni F., Moreira A.: Ultra wide swath imaging with multi-channel ScanSAR. In: Proceedings of IEEE Int Geosci Remote Sens Symp, Boston, Massachusetts 21–24 (2008)
11. Younis M., Bordoni F., Gebert N., Krieger G.: Smart multi-channel radar techniques for spaceborne remote sensing. In: Proceedings of IEEE Int Geosci Remote Sens Symp, Boston, Massachusetts 278–281 (2008)
12. Stiles J., Goodman N., SiChung L.: Performance and processing of SAR satellite clusters. In: Proceedings of IEEE Int Geosci Remote Sens Symp, Honolulu, Hawaii 883–885 (2000)
13. Goodman, N., Lin, S., Rajakrishna, D., Stiles, J.: Processing of multiple-receiver spaceborne arrays for wide-area SAR. IEEE. Trans. Geosci. Remote. Sens. **40**, 841–852 (2002)
14. Aguttes J.P.: The SAR train concept: An along track formation of SAR satellites for diluting the antenna area over N smaller satellites, while increasing performance by N. In: Proceedings of 55th Int Astronautical Congress, Vancouver, Canada 919–925 (2004)
15. Aguttes J.P.: The SAR train concept: required antenna area distributed over N smaller satellites, increase of performance by N. In: Proceedings of IEEE Int Geosci Remote Sens Symp, Toulouse, France 542–544 (2003)
16. Griffiths H., Mancini P.: Ambiguity suppression in SARs using adaptive array techniques. In: Proceedings of IEEE Geoscience and Remote Sensing Symposium, Espoo, Finland 1015–1018 (1991)
17. Callaghan, G.D., Longstaff, I.D.: Wide-swath spaceborne SAR using a quad-element array. IEE. Radar. Sonar. Navig. **146**, 159–165 (1999)

18. Fischer C., Heer C., Krieger G., Werninghaus R.: A high resolution wide swath SAR system. In: Proceedings of European Synthetic Aperture Rdar Conference, Dresden, Germany 1–4 (2006)
19. Li, Z.F., Bao, Z., Wang, H., Liao, G.S.: Performance improvement for constellation SAR using signal processing techniques. IEEE. Trans. Aerosp. Electron. Syst. **42**, 436–452 (2006)
20. Li, Z.F., Bao, Z.: A novel approach for wide-swath and high-resolution SAR image generation from distributed small spaceborne SAR systems. Int. J. Remote. Sens. **27**, 1015–1033 (2006)
21. Jain, A.: Multibeam synthetic aperture radar for global occanography. IEEE. Trans. Antenna. Propag. **27**, 535–538 (1979)
22. Jean, B.R., Rouse, J.W.: A multiple beam synthetic aperture radar design concept for geoscience applications. IEEE. Trans. Geosci. Remote. Sens. **21**, 201–207 (1983)
23. Goodman N., Rajakrishana D., Stiles J.: Wide swath, high resolution SAR using multiple receiver apertures. In: Proceedings of IEEE Int Geosci Remote Sens Symposium, Hamburg, Germany 1767–1769 (1999)
24. Krieger, G., Moreira, A.: Spaceborne bi- and multistatic SAR: potential and challenges. IEE. Radar. Sonar. Navig. **153**, 184–198 (2006)
25. Wang WQ (2010) Bistatic Synthetic Aperture Radar Synchronization Processing. In: Kouemou G (ed) Radar Technology. In-Tech Press, India
26. Wang, W.Q.: Multi-Antenna Synthetic Aperture Radar Imaging: Principles and Applications, In Chinese. National Defense Industry Press, Beijing (2011)
27. Krieger, G., Gebert, N., Moreira, A.: Multidimensional waveform encoding: a new digital beamforming technique for synthetic aperture radar remote sensing. IEEE. Trans. Geosci. Remote. Sens. **46**, 31–46 (2008)
28. Younis M., Venot Y., Wiesbeck J.: Digital beam forming on-receive-only for radar applications. In: Proceedings of German Radar Symposium, Bonn, Germany 213–217 (2002)
29. Younis M.: Digital beam-forming for high resolution wide swath real and synthetic aperture rdar, Dissertation, Karlsruhe, Germany (2004)
30. Krieger G., Fiedler H., Rodriguez-Cassola M., Hounam D., Moreira A.: Analysis of system concepts for bi-and multi-static SAR missions. In: Proceedings of IEEE Geosci Remote Sens Symposium, Toulouse, France 770–772 (2003)
31. Krieger, G., Gebert, N., Moreira, A.: Unambiguous SAR signal reconstruction from nonuniform displaced phase center sampling. IEEE. Geosci. Remote. Sens. Lett. **1**, 260–264 (2004)
32. Krieger G., Gebert N., Moreira A.: SAR signal reconstruction from non-uniform displaced phase center sampling. In: Proceedings of IEEE Int Geosci Remote Sens Symposium, Anchorage, Alaska 1763–1766 (2004)
33. Krieger G., Gebert N., Moreira A. (2004) Digital beamforming and non-uniform displaced phase centre sampling in bi-and multistatic SAR. In: Proc of European Synthetic Aperture Radar Conf, Ulm, Germany 563–566
34. Gebert N., Krieger G., Moreira A.:SAR signal reconstruction from non-uniform displaced phase centre sampling in the presence of perturbations. In: Proceedings of IEEE Int Geosci Remote Sens Symposium, Seoul, Korea 1034–1037 (2005)
35. Gebert N., Krieger G., Moreira A.: High resolution wide swath SAR imaging with digital beamforming-performance analysis, optimization and system design. In: Proceedings of European Synthetic Aperture Radar Conference, Dresden, Germany 341–344 (2006)
36. Gebert N., Krieger G., Moreira A.: Digital beamforming for HRWS-SAR imaging system design, performance and optimization strategies. In: Proceedings of IEEE Int Geosci Remote Sens Symposium, Denver, Colorado 1836–1839 (2006)
37. Gebert N., Krieger G., Moreira A.: Multi-channel ScanSAR for high-resolution ultra-wide-swath imaging. In: Proceedings of European Synthetic Aperture Radar Conference, Friedrichshafen, Germany 79–82 (2008)
38. Claassen J.P., Eckerman J.: A system concept for wide swath constant incident angle coverage. In: Proceedings of Synthetic Aperture Radar Technology Conference, Las Cruces, New Mexico 41–59 (1978)

39. Bellettini, A., Pinto, M.A.: Theoretical accuracy of synthetic aperture sonar micronavigation using a displaced phase-center antenna. IEEE. J. Oceanic. Enginneer. **27**, 780–789 (2002)
40. Lombardo, P., Colone, F., Pastina, D.: Monitoring and surveillance potentialities obtained by splitting the antenna of the COSMO-SkyMed SAR into multiple sub-apertures. IEE. Proc. Radar. Sonar. Navig. **153**, 104–116 (2006)
41. Wang, W.Q., Cai, J.Y., Peng, Q.C.: Conceptual design of near-space synthetic aperture radar for high-resolution and wide-swath imaging. Aerosp. Sci. Technol. **13**, 340–347 (2009)
42. Cumming, I.G., Wong, F.H.: Digital Processing of Synthetic Aperture Radar Data. Artech House, Boston (2005)
43. Li, F.K., Johnson, W.T.K.: Ambiguities in spaceborne synthetic aperture radar systems. IEEE. Trans. Aerosp. Electron. Syst. **19**, 389–397 (1983)
44. Krieger G., Moreira A.: Potentials of digital beamforming in bi-and multistatic SAR. In: Proceedings of IEEE Geosci Remote Sens Symposium, Toulouse, France 527–529 (2003)
45. Younis, M., Fischer, C., Wiesbeck, W.: Digital beamforming in SAR systems. IEEE. Trans. Geosci. Remote. Sens. **41**, 1735–1739 (2003)
46. Gebert N., Krieger G., Moreira A.: Digital beamforming for HRWS-SAR imaging. In: Proceedings of IEEE Geosci Remote Sens Symposium, Denver, USA 1836–1839 (2006)
47. Gebert, N., Krieger, G., Moreira, A.: Digital beamforming on receive: techniques and optimization stragies for high-resolution wide-swath SAR imaging. IEEE. Trans. Aerosp. Electron. Syst. **45**, 564–592 (2009)
48. Younis M., Patyuchenko A., Huber S., Bordoni F., Krieger G.: Performance comparison of reflector-and planar-antenna based digital beam-forming SAR. Int J Antenna Propag (2010) doi:10.1155/2009/614931
49. Huber S., Younis M., Patyuchenko A., Krieger G.: Digital beam forming techniques for spaceborne reflector SAR systems. In: Proceedings of European Synthetic Aperture Radar Conference, Aachen, Germany 962–965 (2010)
50. Wang, W.Q., Peng, Q.C., Cai, J.Y.: Waveform-diversity-based millimeter-wave UAV remote sensing. IEEE. Trans. Geosci. Remote. Sens. **47**, 691–700 (2009)
51. Wang, X.Q., Xiao, Q., Chen, Y.Q., Zhu, M.H.: The SNR study of the wide-swath SAR basing on elevation multi-receiver. J. Electron. Info. Technol. **29**, 2101–2104 (2007)
52. Krieger, G., Gebert, N., Moreira, A.: Multidimensional waveform encoding: a new digital beamforming technique for synthetic aperture radar remote sensing. IEEE. Trans. Geosci. Remote. Sens. **46**, 31–45 (2008)
53. Wang W.Q.: Applications of MIMO technique for aerospace remote sensing. In: Proceedings of IEEE Aerospace Conference, Big Sky, MT 1–10 (2007)
54. Zhuge, X.D., Yarovoy, A.G.: A sparse aperture MIMO-SAR-based UWB imaging systems for concealed weapon detection. IEEE. Trans. Geosci. Remote. Sens. **49**, 509–518 (2011)
55. Cristallini, D., Pastina, D., Lombardo, P.: Exploiting MIMO SAR potentialities with efficient cross-track constellation configurations for improved range resolution. IEEE. Trans. Geosci. Remote. Sens. **49**, 38–52 (2011)
56. Wang W.Q.: Space-time coding MIMO-OFDM SAR for high-resolution imaging. IEEE. Trans. Geosci. Remote. Sens. (2011) doi:10.1109/TGRS.2011.2116030
57. Picciolo M.S., Griesbach J.D., Gerlach K.: Adaptive LFM waveform diversity. In: Proceedings of IEEE Radar Conference, Rome, Italy, 1–6 (2008)
58. Levanon, N., Mozeson, E.: Radar Signals. Wiley-IEEE Press, New York (2004)
59. Kim J.H., Ossowska A., Wiesbeck W.: Investigation of MIMO SAR for interferometry. In: Proceedings of 4th European Radar Conference, Munich Germany 51–54 (2007)

Chapter 5
Near-Space Vehicles in Ground Moving Target Indication

Abstract The last but not the least application of near-space vehicles in remote sensing is the Ground moving target indication (GMTI). In this chapter, we aim at MIMO SAR-based solution for GMTI including image formation processing. GMTI using MIMO SAR is interesting because the target indication performance can be improved by MIMO SARs' larger virtual aperture. The minimum detectable velocity of a target can be improved by both the larger virtual aperture size and the longer coherent processing interval used by MIMO SARs.

Keywords Near-space · Ground moving target indication (GMTI) · MIMO SAR · Along-track interferometry (ATI) · Multi-antenna SAR · Fractional Fourier transform (FrFT)

5.1 MIMO SAR with Multi-Antenna in Azimuth

GMTI is of great important for surveillance and reconnaissance [1–6]. One representative GMTI technique is the along-track interferometry (ATI) SAR, which was initially proposed for estimating the radial velocity of ground moving targets, starting from interferometric phase measurements [7–9]. However, ATI SAR does not take into account the fact that the stationary clutter unavoidably corrupts the interferometric phase of the target depending on its signal-to-clutter environment [10]. Consequently the imaged moving targets may be displaced in azimuth according to its radial velocity and superimposed upon clutter at a wrong location. Moreover, the estimated velocity may be ambiguous. To overcome this disadvantage, a method of moving target detection and location with three-frequency three-aperture ATI SAR was proposed in [11], but the parameter estimation and image formation algorithms are not discussed. Differently, here we discuss MIMO SAR-based GMTI approaches.

The multiple antennas in MIMO SAR systems can be placed in elevation and/or azimuth, but we consider only multiple antennas in azimuth. Figure 5.1 shows the geometry of the MIMO SAR with multi-antenna in azimuth for GMTI applications.

Fig. 5.1 Geometry of
MIMO SAR with
multi-antenna in azimuth

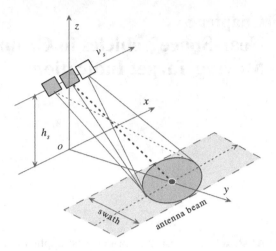

The SAR platform is moving along x-axis at a constant speed of v_s and at an altitude of h_s. Suppose the transmit array has N_t co-located antennas whereas the receive array has N_r co-located antennas. The waveforms used in any two transmit antennas must be orthogonal, so that their echoes can be separated in subsequent signal processing. For each of the N_t transmitted signals, there are N_r coherent returns; hence, there will be $N_t \times N_r$ different returns, each is weighed by different amplitude and phase. Each receive antenna signal is separately demodulated, digitized, and stored. Next, multiple pairs of equivalent transmitter-to-receiver phase centers can be formed by a jointly spatiotemporal processing. This provides a potential solution to GMTI processing.

Also, we make the assumptions of far-field, flat earth, free-space, and single-polarization for our model. Suppose there are three along-track antennas. The range history from the central antenna to a given moving target located at (x_0, y_0) with a velocity of (v_x, v_y)(v_x and v_y denote the velocity elements in x-axis and y-axis, respectively) is represented by

$$R_2(\tau) = \sqrt{[x_0 + (v_x - v_s)\tau]^2 + (y_0 + v_y\tau)^2 + h_s^2}$$
$$\approx R_0 + v_y\tau + \frac{[x_0 + (v_s - v_x)\tau]^2 + (v_y\tau)^2}{2R_0} \tag{5.1}$$

where τ is the azimuth slow time. The reference range R_0 is

$$R_0 = \sqrt{y_0^2 + h_s^2}. \tag{5.2}$$

Similarly, the range histories from the target to the left antenna and right antenna are represented, respectively, by

$$R_1(\tau) \approx R_0 + v_y\tau + \frac{[x_0 + d_a + (v_s - v_x)\tau]^2 + (v_y\tau)^2}{2R_0} \tag{5.3}$$

$$R_3(\tau) \approx R_0 + v_y\tau + \frac{[x_0 - d_a + (v_s - v_x)\tau]^2 + (v_y\tau)^2}{2R_0} \tag{5.4}$$

where d_a is the separation distance between two neighboring antennas in azimuth. Without loss of generality, amplitude terms are ignored in the following discussions. Suppose $s_m(t)$ is the signal transmitted by the mth ($m \in (1, 2, 3)$) antenna with carrier frequency $f_{c,m}$, the echoes received by the nth ($n \in (1, 2, 3)$) receive antenna can then be expressed as

$$\begin{aligned} r_{m,n}(t, \tau) = s_m &\left(t - \frac{R_m(\tau) + R_n(\tau)}{c_0}\right) \\ &\times \exp\left\{j2\pi f_c\left[t - \frac{R_m(\tau) + R_n(\tau)}{c_0}\right]\right\} \end{aligned} \tag{5.5}$$

where t and f_c denote the range fast time and radar carrier frequency, respectively. As the transmitted waveforms are orthogonal, after separately range pulse compression, we have

$$s_{n,m}(t, \tau) = \exp\left(-j2\pi f_{c,m}\tau_{n,m}\right) \cdot \mathrm{sinc}[\pi B_r(t - \tau_{n,m})] \tag{5.6}$$

where $\tau_{n,m}$ is the signal propagation time from the mth transmit antenna to the nth receive antenna

$$\tau_{n,m} = \frac{R_m(\tau) + R_n(\tau)}{c_0}. \tag{5.7}$$

The term $s_{n,m}(t, \tau)$ denotes the matched filtering output at the nth receive antenna and matched to the mth transmit antenna waveform. This is just the signal model of the MIMO SAR with multi-antenna in azimuth, which means that there are a total of 3×3 matched filter outputs (or sub-images) resulting from a spatial convolution of the transmit and receive arrays.

As the multiple antennas are displaced in azimuth direction, it is effective to define "two-way" phase center as the mid-point between the transmit and receive phase centers. This provide a potential clutter suppression, like the DPCA technique. The clutter cancelation can be performed by subtracting the samples of the returns received by two-way phase centers located in the same spatial position, which are temporally displaced. The returns corresponding to stationary objects like the clutter coming from natural scene are cancelled, while the returns backscattered from moving targets have a different phase center in the two acquisitions and remain uncanceled. Therefore, all static clutter are cancelled, leaving only moving targets.

5.2 MIMO SAR-Based GMTI Processing

Consider the three azimuth-antennas shown in Fig. 5.2. For the mth transmit antenna, the range pulse compression results at the three receive antennas are represented, respectively, by

Fig. 5.2 Scheme of MIMO
SAR with three
azimuth-antennas for GMTI
applications

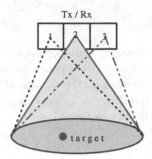

$$s_{1,m}(t, \tau) = \exp\left(-j2\pi f_{c,m}\tau_{1,m}\right) \cdot \mathrm{sinc}[\pi B_r(t - \tau_{1,m})] \qquad (5.8\mathrm{a})$$

$$s_{2,m}(t, \tau) = \exp\left(-j2\pi f_{c,m}\tau_{2,m}\right) \cdot \mathrm{sinc}[\pi B_r(t - \tau_{2,m})] \qquad (5.8\mathrm{b})$$

$$s_{3,m}(t, \tau) = \exp\left(-j2\pi f_{c,m}\tau_{3,m}\right) \cdot \mathrm{sinc}[\pi B_r(t - \tau_{3,m})]. \qquad (5.8\mathrm{c})$$

To cancel stationary clutter, we perform DPCA processing

$$sc_{m,21}(t, \tau) = s_{2,m}(t, \tau) - s_{1,m}(t, \tau + \Delta\tau) \cdot G_0 \qquad (5.9)$$

$$sc_{m,32}(t, \tau) = s_{3,m}(t, \tau - \Delta\tau) \cdot G_0 - s_{2,m}(t, \tau) \qquad (5.10)$$

where $\Delta\tau = d_a/(2v_s)$ is the relative azimuth time delay between two neighboring
antennas, and G_0 is used to compensate the corresponding phase shift between two
antennas

$$G_0 = \exp\left(-j\frac{2\pi d_a^2}{4R_0\lambda}\right). \qquad (5.11)$$

Since there is

$$A_0 = \mathrm{sinc}[\pi B_r(t - \tau_{1,m})] \approx \mathrm{sinc}[\pi B_r(t - \tau_{2,m})] \approx \mathrm{sinc}[\pi B_r(t - \tau_{3,m})]. \quad (5.12)$$

The Eqs. 5.9 and 5.10 can be derived into

$$sc_{m,21}(t, \tau) = A_0 \cdot \exp\left(-j2\pi f_{c,m}\tau_{2,m}\right) \cdot \exp\left(-j\frac{2\pi}{\lambda_m}v_y\Delta\tau\right)$$
$$\cdot \left[1 - \exp\left(-j\frac{2\pi}{\lambda_m}v_y\Delta\tau\right)\right] \qquad (5.13)$$

$$sc_{m,32}(t, \tau) = A_0 \cdot \exp\left(-j2\pi f_{c,m}\tau_{2,m}\right) \cdot \exp\left(j\frac{2\pi}{\lambda_m}v_y\Delta\tau\right) \cdot \left[1 - \exp\left(-j\frac{2\pi}{\lambda_m}v_y\Delta\tau\right)\right]$$
$$(5.14)$$

where λ_m is the wavelength. It can be noticed that, if $v_y = 0$, there will be $|sc_{m,21}(t, \tau)$ $= 0|$ and $|sc_{m,32}(t, \tau) = 0|$; hence, the clutter is successfully canceled by this method. The next problem is to detect the moving targets.

To estimate the moving target's radial velocity v_y, we perform

$$sc_{m,32-21}(t, \tau) = \frac{sc_{m,32}(t, \tau)}{sc_{m,21}(t, \tau)} = \exp\left(j\frac{4\pi}{\lambda_m}v_y\Delta\tau\right). \tag{5.15}$$

We then have

$$sc_{1,32-21}(t, \tau) = \exp\left(j\frac{4\pi}{\lambda_1}v_y\Delta\tau\right) \tag{5.16a}$$

$$sc_{2,32-21}(t, \tau) = \exp\left(j\frac{4\pi}{\lambda_2}v_y\Delta\tau\right) \tag{5.16b}$$

$$sc_{3,32-21}(t, \tau) = \exp\left(j\frac{4\pi}{\lambda_3}v_y\Delta\tau\right). \tag{5.16c}$$

Next, we can get the following interferometry phases

$$\Delta\Phi_1 = \arg\{sc_{2,32-21}(t, \tau) \cdot sc^*_{1,32-21}(t, \tau + \Delta\tau)\}$$

$$= \exp\left[j4\pi v_y\Delta\tau\left(\frac{1}{\lambda_1} - \frac{1}{\lambda_2}\right)\right] \tag{5.17}$$

$$\Delta\Phi_2 = \arg\{sc_{3,32-21}(t, \tau) \cdot sc^*_{2,32-21}(t, \tau + \Delta\tau)\}$$

$$= \exp\left[j4\pi v_y\Delta\tau\left(\frac{1}{\lambda_2} - \frac{1}{\lambda_3}\right)\right] \tag{5.18}$$

$$\Delta\Phi_3 = \arg\{sc_{3,32-21}(t, \tau) \cdot sc^*_{1,32-21}(t, \tau + 2\Delta\tau)\}$$

$$= \exp\left[j4\pi v_y\Delta\tau\left(\frac{1}{\lambda_1} - \frac{1}{\lambda_3}\right)\right]. \tag{5.19}$$

The radial velocity v_y can then be calculated from Eqs. 5.17 to 5.19. Taking Eq. 5.19 as an example, we can get

$$v_y = \frac{v_s\Delta\Phi_3\lambda_1\lambda_3}{4\pi v_s\Delta\tau(\lambda_1 - \lambda_3)} = \frac{v_s\Delta\Phi_3\lambda_1\lambda_3}{2\pi d(\lambda_1 - \lambda_3)}. \tag{5.20}$$

The corresponding unambiguous speed estimation range $v_{y_{um}}$ is determined by

$$-\frac{v_s\lambda_3\lambda_1}{2d_a(\lambda_1 - \lambda_3)} \leq v_{y_{um}} \leq \frac{v_s\lambda_3\lambda_1}{2d_a(\lambda_1 - \lambda_3)}. \tag{5.21}$$

It can be noticed that the unambiguous speed estimation range has been significantly extended when compared to the conventional single-carrier detection method. This is particularly valuable for GMTI applications. However, it is necessary to be noted that, to avoid possible blind speed problem in target detection, all the three interferometry phases should be used to calculate the radial speed. This means that we can get possible three speeds and the true speed is identified with jointly processing, so as to avoid loss detection of the moving targets.

The double-interferometry method expressed in Eq. 5.20 has a good unambiguous speed estimate range which is an advantage for detecting high-speed targets, but its estimate precision may be unsatisfactory for some specific applications. To overcome this disadvantage, we firstly use the speed obtained from the previous double-interferometry method to de-ambiguity the targets. Next, after range mitigation correction we can perform interferometry processing in range-Doppler domain for the same transmit antenna [11]

$$\Delta\Phi'_m = \arg\{sc_{m,32}(r, f_d) \cdot sc^*_{m,21}(r, f_d)\} = \frac{2\pi v_{y_e} d_a}{\lambda_m v_s}. \tag{5.22}$$

We can then obtain the residual radial speed

$$v_{y_e} = \frac{\lambda_m v_s \Delta\Phi'_m}{2\pi d_a}. \tag{5.23}$$

Therefore, both unambiguous speed estimate range and precision can be improved by this jointly processing algorithm.

To estimate the moving targets' Doppler parameters, we transform the range equation expressed in Eq. 5.1 into

$$R_2(\tau) \approx R_0 + \frac{x_0^2}{2R_0} + \frac{y_0 v_y + x_0(v_x - v_s)}{R_0}\tau + \frac{v_y^2 + (v_x - v_s)^2}{2R_0}\tau^2, \tag{5.24}$$

and substitute it into Eq. 5.13 which yields

$$s_c(\tau) = A_1 \exp[-j\pi(k_d\tau^2 + 2f_{dc}\tau) + \phi_0] \tag{5.25}$$

with

$$A_1 = A_0 \exp\left(-j\frac{2\pi}{\lambda_m}v_y\Delta\tau\right) \cdot \left[1 - \exp\left(-j\frac{2\pi}{\lambda_m}v_y\Delta\tau\right)\right] \tag{5.26a}$$

$$k_d = \frac{2}{\lambda_2 R_0}[(v_s - v_x)^2 + v_y^2] \tag{5.26b}$$

$$f_{dc} = \frac{2}{\lambda_2 R_0}[y_0 v_y + x_0(v_x - v_s)] \tag{5.26c}$$

$$\phi_0 = -\frac{4\pi}{\lambda}\left(R_0 + \frac{x_0^2}{2R_0}\right). \tag{5.26d}$$

Therefore, once the Doppler parameters k_d and f_{dc} are obtained with estimation algorithms (detailed in next section), the parameters of v_x, y_0, and R_0 can be obtained by the Eqs. 5.2 and 5.26 because the h_s in Eq. 5.2 is knowable from the inboard motion measurement or compensation sensors.

5.3 Simplified FrFT-Based Parameters Estimation

To estimate the Doppler parameters k_d and f_{dc}, we developed a simplified fractional Fourier transform (FrFT)-Based algorithm.

5.3.1 Simplified FrFT Algorithm

The conventional FrFT is a hybridized time-frequency transform algorithm. Its transform kernel is defined as [12]

$$
K_\alpha(t, \mu) = \begin{cases} \sqrt{\dfrac{1-j\cot\alpha}{2\pi}} e^{j\frac{t^2+\mu^2}{2}\cot\alpha - j\mu t \csc\alpha}, & \alpha \neq k\pi \\ \delta(t-\mu), & \alpha = 2k\pi \\ \delta(t+\mu), & \alpha = (2k+1)\pi \end{cases} \tag{5.27}
$$

where k denotes an integer, and α indicates the rotation angle in FrFT domain. This operation can be considered as a generalized form of Fourier transform that corresponds to a rotation over an arbitrary angle $\alpha = r\pi/2$ with $r \in \Re$. The forward and inverse FrFT of $x(t)$ are defined, respectively, by

$$
\chi_\alpha(\mu) = \int_{-\infty}^{\infty} x(t) K_\alpha(t, \mu) \, dt \tag{5.28}
$$

$$
x(t) = \int_{-\infty}^{\infty} \chi_\alpha(\mu) K_{-\alpha}(t, \mu) \, du. \tag{5.29}
$$

The FrFT of a function $x(t)$, with an angle α, can be computed in the following steps.

Step 1. A product by a chirp:

$$
g(t) = x(t) e^{-j\frac{1}{2}t^2 \tan(\frac{\alpha}{2})}. \tag{5.30}
$$

Step 2. A Fourier transform (with its argument scaled by $\csc(\alpha)$) or a convolution:

$$
g^*(t) = g(t) \circledast e^{-j\frac{1}{2}t^2 \csc(\alpha)} = \int_{-\infty}^{\infty} x(t) e^{-j\frac{1}{2}(\mu-t)^2 \csc(\alpha)} \, dt \tag{5.31}
$$

where $\circledast$ denotes a convolution operator.

Step 3. Another product by a chirp:

$$f(t) = e^{-j\frac{1}{2}\mu^2 \tan(\frac{\alpha}{2})} g^*(t). \tag{5.32}$$

Step 4. A product by a complex amplitude factor:

$$F_r^\alpha[x(t)] = \sqrt{\left(\frac{1 - j\cot(\alpha)}{2\pi}\right)} f(t). \tag{5.33}$$

As the Steps 3 and 4 are redundant for signal detection, we name the FrFT without the Steps 3 and 4 as the simplified FrFT (sFrFT) [13]. It is also a linear transform and continuous in the angle α. This paper just uses this sFrFT to estimate the Doppler parameters.

5.3.2 Simplified FrFT-Based Estimation Algorithm

To estimate the Doppler parameters, applying the sFrFT to the Eq. 5.25, we get

$$\chi_\alpha(\mu) = A'_o \exp(j\phi_1)$$
$$\cdot \int_{-T_p/2}^{T_p/2} \exp\left[j2\pi f_{dc}\tau + j\pi k_d\tau^2 + j\frac{\tau^2}{2}\cot(\alpha) - j\mu\tau\csc(\alpha) + j\phi_1\right]d\tau. \tag{5.34}$$

It arrives its maximum at [14]

$$\{\hat{\alpha}_0, \hat{\mu}_0\} = \arg\max_{\alpha,\mu} |X_p(\mu)|^2 \tag{5.35}$$

$$\begin{cases} \hat{k}_d = -\cot(\hat{\alpha}_0) \\ \hat{\alpha}_0 = \frac{|X_{\hat{p}}(\hat{\mu}_0)|}{\Delta\tau} \\ \hat{f}_{dc} = \hat{\mu}_0 \csc(\hat{\alpha}_0). \end{cases} \tag{5.36}$$

This condition forms the basis of estimating the moving targets' parameters. In the sFrFT domain with a proper α, the spectra of any strong moving target will concentrate to a narrow impulse, and that of the clutter will be spread. If we can construct a narrow band-stop filter in the sFrFT domain whose center frequency around at the center of the narrowband spectrum of a strong moving target, then the signal component of this moving target can be extracted from the initial signal. With this method the strong moving targets can be extracted iteratively, thereafter the weak moving targets may be detectable. This method can be regarded as an extension of the CLEAN algorithm [15] to the sFrFT.

Therefore, after canceling the stationary clutter, the identification of the moving targets can be implemented with sFrFT in the following steps:

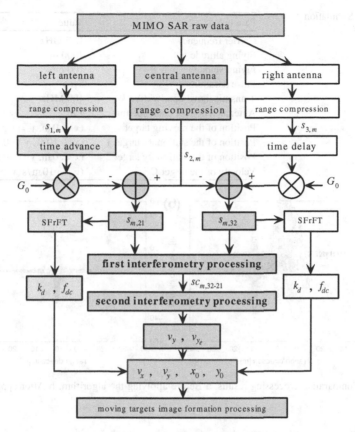

Fig. 5.3 The flow chart of the sFrFT and double-interferometry processing combined algorithm

Step 1. Apply one sFrFT to the data in which the clutter has been canceled with different α, and from the maximal peak we can get the numerical estimation of $(\hat{\mu}, \hat{\alpha})$.

Step 2. Apply $F_r^{\hat{\alpha}}$ to the same data, we have

$$X_{\hat{\alpha}}(\mu) = \chi_{\hat{\alpha}}(\mu). \qquad (5.37)$$

Step 3. After identifying the first moving target, we then construct a narrow band-stop filter $M(\mu)$ to notch the narrow band-stop spectrum of this moving target

$$X'_{\hat{\alpha}}(\mu) = X_{\hat{\alpha}}(\mu)M(\mu). \qquad (5.38)$$

Step 4. The filtered signals are then rotated back to time-domain by an inverse sFrFT.

Step 5. Repeat the operations from Step 1 to Step 4 until all the desired moving targets are identified.

Table 5.1 Simulation parameters

Parameters	Values
Carrier frequency	1.25 GHz
Flying altitude	4000 m
Flying velocity	100 m/s
Pulse duration	$5\,\mu s$
Transmit signal bandwidth	100 MHz
Antenna length of each aperture	1 m
Position of the moving target	$(x = 0\,m, y = 0\,m)$
Position of the stationary target 1	$(x = 10\,m, y = 0\,m)$
Position of the stationary target 2	$(x = 0\,m, y = 10\,m)$
Velocity of the target C	$(v_x = 10\,m/s, v_y = 5\,m/s)$

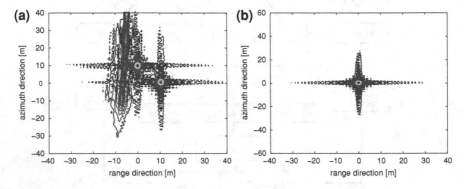

Fig. 5.4 Comparative processing results. **a** Before applying the algorithm. **b** After applying the algorithm

Once the Doppler parameters of each target are obtained, we can get the v_x. Now, the parameters (v_x, v_y) and (x_o, y_o) are all determined successfully. Next, the moving targets can then be focused with one uniform image formation algorithm, such as range-Doppler (RD) and chirp scaling (CS) algorithms. An example processing steps are given in Fig. 5.3.

5.4 Simulation Results

Unmanned vehicle-borne three-antenna stripmap MIMO SAR data from three point targets, one moving target and two stationary targets, are simulated using the parameters listed in Table 5.1. Figure 5.4 shows the comparative processing results using the CS-based image formation processing algorithm. It can be noticed that the imaged moving target is overlapped with the two stationary targets. The moving target cannot be identified from this figure, because we cannot discern which is the moving one and which is the stationary one. Next, clutter cancellation is obtained by applying the proposed double-interferometry processing technique, leaving only moving targets and a much simplified target detection problem. The sFrFT algorithm is then applied to estimate the Doppler parameters. Finally, the moving targets are focused by the image formation algorithm with the estimated Doppler parameters.

5.5 Conclusion

In this chapter, we proposed a scheme of MIMO SAR for ground moving targets detection and imaging. Although MIMO radar has received considerable attention in recent years, little work is related to MIMO SAR remote sensing. Our approach employs MIMO along-track antenna configuration and waveform diversity. After cancelling clutter and estimating the along-track velocity of the moving targets with the proposed double-interferometry processing algorithm and estimating the Doppler parameters of the moving targets with the presented sFrFT technique, GMTI is obtained by a jointly processing algorithm. Finally, the moving targets are focused with one uniform image formation algorithm. In this way, both target location and target velocity can be acquired, and high-resolution moving targets' SAR imagery can also be obtained. Simulation results show its validity.

As the conventional DPCA SAR detection performance is noise limited and the conventional ATI SAR detection is clutter limited, the proposed approach combines their advantages and overcomes their disadvantages; hence, this approach is more effective and robust than the conventional DPCA SAR-based GMTI approaches. In particular, it is not dependent on a target's across-track velocity component or its Doppler shift, which is difficult to determine due to insufficient freedom degrees. Moreover, it also allows the estimation of target's true azimuth position directly from its measured position in the final SAR images. Therefore, the MIMO SAR and double-interferometry processing combined GMTI approach is elegant and effective in moving target identification.

References

1. Gierull, C.H., Maori, D.C., Ender, J.: Ground moving target indication with tandem satellite constellations. IEEE. Geosci. Remote. Sens. Lett. **5**, 710–714 (2008)
2. Zhu, S.Q., Liao, G.S., Qu, Y., Liu, X.Y., Zhou, Z.G.: A new slant-range velocity ambiguity resolving approach of fast moving targets for SAR system. IEEE. Trans. Geosci. Remote. Sens. **48**, 432–451 (2010)
3. Park, S.H., Kim, H.T., Kim, K.T.: Segmentation of ISAR images of targets moving in formation. IEEE. Trans. Geosci. Remote. Sens. **48**, 2099–2108 (2010)
4. Li, G., Xu, Y.N., Peng, Y.N., Xia, X.G.: Bistatic linear antenna array SAR for moving target detection, location and imaging with two passive airborne radars. IEEE. Trans. Geosci. Remote. Sens. **45**, 554–565 (2007)
5. Wang, G., Xia, X.G., Chen, V.C., Fiedler, R.L.: Detection, location, and imaging fast moving targets using multifrequency antenna array SAR. IEEE. Trans. Aerosp. Electron. Syst. **40**, 345–355 (2004)

6. Li, G., Xia, X.G., Peng, Y.N.: Doppler keystone transform: an approach suitable for parallel implementation of SAR moving target imaging. IEEE. Geosci. Remote. Sens. Lett. **5**, 573–577 (2008)
7. Budillon, A., Pascazio, V., Schirinzi, G.: Estimation of radial velocity of moving targets by along-track interferometric SAR systems. IEEE. Geosci. Remote. Sens. Lett. **5**, 349–353 (2008)
8. Yang, L., Wang, T., Bao, Z.: Ground moving target indication using an InSAR system with hybrid baseline. IEEE. Geosci. Remote. Sens. Lett. **5**, 373–377 (2008)
9. Guarbieri, A.M., Tebaldini, S.: On the exploitation of target statistics for SAR interferometry applications. IEEE. Trans. Geosci. Remote. Sens. **46**, 3436–3443 (2008)
10. Moya, J.C., Menoyo, J.G., Lopez, A.A., Blance-del-Campo, A.: Application f the random transform to detect small-targets in sea clutter. IET. Rdar. Sonar. Navig. **3**, 155–166 (2009)
11. Yin, J.F., Li, D.J., Wu, Y.R.: Research on the method of moving target detection and location with three-frequency three-aperture along-track spaceborne SAR. J. Electron. Info. Technol. **32**, 902–907 (2010)
12. Almeida, L.B.: The fractional Fourier transform and time-frequency representations. IEEE. Trans. Sig. Process. **42**, 3084–3091 (1994)
13. Wang WQ (2005) Estimating the doppler centroid of moving targets for spaceborne SAR based on fractional Fourier transform. In: Proceedings of International Symposium Physical Meas Signatures in Remote Sens, Beijing, China, pp. 17–20
14. Wang, W.Q.: Approach of multiple moving targets detection for microwave surveillance sensors. Int. J. Infor. Acquisition. **4**, 57–68 (2007)
15. Tsao, J., Steinber, B.D.: Reduction of sidelobe and speckle artifacts in microwave imaging. IEEE Trans Antenna Propag **36**, 543–556 (1988)

Chapter 6
Summary

Abstract Near-space does indeed seem to be an efficient solution to future remote
sensing applications. The use of cost-effective near-space platforms can lead to the
solutions that previously thought to be out of reach for remote sensing customers.
However, there are several technological challenges. In this chapter, we discuss realistic near-space remote sensing issues and possible future work.

Keywords Near-space · Near-space remote sensing · High-precision imaging ·
Waveform diversity · Three-dimensional imaging · Near-space vehicle

6.1 Realistic Near-Space Remote Sensing Issues

Weather could be a significant problem if near-space vehicles are not furnished
with reliable sensors for on-site meteorological data with which vehicle controllers
can predict turbulence, icing, and violent gusts. The experience with high-altitude
tropospheric operations from around-the-world balloonist teams and weather teams
could be collected and codified to aid computer predictions at higher altitudes.

Another limitation is the difficulty in getting vehicles to altitudes higher than
just the lower reaches of near-space. Increasing vehicle altitudes greatly increases
vehicle balloon size, quickly making them too difficult to manage on the ground and
during launch. A rule of thumb is that for every extra 30 km of altitude to reach,
the vehicle needs to have twice the volume. Vehicles quickly become unmanageable
for repetitive and hasty military use if we try to consistently reach high altitudes.
This is perhaps why the Air Force Scientific Advisory Board recently considered
near-space vehicles higher than 30 km "not viable" in the near future. However,
this unfortunately leaves near-space vehicles in the altitude region that will probably
make them more susceptible to enemy attack.

Another possible problem may be the loss of tactical or operational surprise.
First, low-tech adversaries, such as terrorists and third world countries, may not

W.-Q. Wang, *Near-Space Remote Sensing*,   111
SpringerBriefs in Electrical and Computer Engineering,
DOI: 10.1007/978-3-642-22188-0_6, © Wen-Qin Wang 2011

have the capability to easily track or monitor the vehicles. Also, even radar-capable adversaries may have difficulty in tracking vehicles at longer distances due to their slow speed and small RCS. Secondly, near-space vehicles will be loiter for a long time, and possibly be launched with such repetition that their existence will not provide the enemy any valuable intelligence.

6.2 Future Work

There are key technological issues which are still under research and require further development. A brief overview of issues related with near-space remote sensing is given here.

6.2.1 High-Precision Imaging Algorithm

One future work is to develop high-precision image formation algorithms for near-space vehicle-borne azimuth-variant BiSAR and multi-antenna SAR systems. For near-space vehicle-borne azimuth-variant BiSAR it is a mandatory to know both the transmitter-to-receiver distance and target-to-receiver distance to properly focus its raw data, but they clearly depend on the target height. In this case, conventional imaging algorithms may be not suitable to accurately focus the collected data. Azimuth-variant BiSAR data can be focused in time-domain and frequency-domain or hybrid-domain. As high-computational cost is required for time-domain processing algorithms, the frequency-domain or hybrid-domain processing algorithms are of great interest. However, their processing efficiency depend on the spectrum model of the echo signal. One equivalent velocity-based two-dimensional spectrum model was proposed in Chap. 3, but only the response of a point target was simulated. We are aware that much further work is required to develop practical algorithms for distributed or multiple targets imaging. Also, near-space vehicle-borne multi-antenna SAR imaging algorithms requires much further work. Since multi-antenna SAR is in its infancy, there is a lack of efficient imaging algorithms for them.

6.2.2 Waveform Diversity Design

Continuous frequency coverage in the OFDM-LFM waveform is assumed in Chap. 4; however, if there are frequency overlaps or frequency separations in the waveform, unwanted side lobes will be generated in subsequent digital beamforming processing. Figure 6.1 shows the impacts of frequency overlap and separation on final matched filtering results. A frequency overlap of $0.5B$ and a frequency separation of $0.5B$ are assumed, respectively. It can be noticed that the frequency

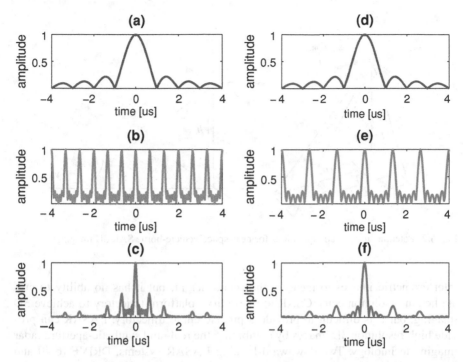

Fig. 6.1 Impacts of frequency overlap and separation on final matched filtering results: **a–c** the case with frequency overlap, **d–f** the case with frequency separation

overlap and separation interval between the sub-chirp signals play an important role on the sidelobe performance. When continuous frequency coverage is used, a good sidelobe performance can be obtained. Otherwise, the sidelobe performance will be degraded. To further suppress the sidelobes, some other efficient algorithms should be developed.

6.2.3 Three-Dimensional Imaging

Another future work is near-space vehicle-borne three-dimensional (3D) imaging. Conventional SAR obtains its two-dimensional (2D) high-resolution images along azimuth and elevation dimensions by projecting the 3D distributed targets onto the 2D plane. Consequently it usually suffers from geometric distortions, such as foreshortening and layover. The strong shadowing effects caused by building, hills, and valleys may result in the information loss of the explored area. In order to overcome these disadvantages, 3D SAR imaging has become an urgent need [1]. Current 3D SAR imaging techniques include interferometry SAR (InSAR), curved SAR (CSAR) [2–4], and linear array SAR (LASAR) [5]. InSAR can create

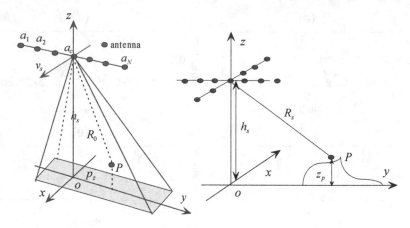

Fig. 6.2 Potential antenna configurations for near-space vehicle-borne SAR 3D imaging

interferometric images to measure the scene height, but it has no ability to create height resolution cells. CSAR utilizes curve platform trajectory to achieve 3D imaging, but it is difficult to provide a precise curve trajectory. LASAR can provide high-resolution 3D images by combining the real- and synthetic-aperture radar imaging technology. Two downward-looking LASAR systems, DRIVE [6–8] and ARTINO [9–11], are being developed at ONERA and FGAN-FHR, respectively. However, the antenna arrays used in these two systems are SIMO (single-input and multiple-output) configuration. It is necessary to extend it into MIMO configuration. Figure 6.2 shows two potential configurations for near-space vehicle-borne SAR 3D imaging.

References

1. Fornaro, G., Lombardini, F., Serafino, F.: Three-dimensional multipass SAR focusing: experiments with long-term spaceborne data. IEEE Trans. Geosci. Remote Sens. **43**, 702–714 (2005)
2. Ishimaru, A., King, C.T., Kuga, Y.: An imaging technique using confocal circular synthetic aperture radar. IEEE Trans. Geosci. Remote Sens. **36**, 1524–1530 (1998)
3. King, C.T., Kuga, Y., Ishimaru, A.: Experimental studies on circular SAR imaging in clutter using angular correlation function technique. IEEE Trans. Geosci. Remote Sens. **37**, 2192–2197 (1999)
4. Axelsson, S.R.J.: Beam characteristics of three-dimensional SAR in curved or random paths. IEEE Trans. Geosci. Remote Sens. **42**, 2324–2334 (2004)
5. Mahafza, B.R., Sajjadi, M.: Three-dimensional SAR imaging using linear array in transverse motion. IEEE Trans. Aerosp. Electron. Syst. **22**, 499–510 (1996)
6. Nouvel, J.F., Jeuland, H., Bonin, G., Roques, S., du Plessis, O., Peyret, J.: A Ka band imaging radar DRIVE on board ONERA motorglider. In: Proceedings of IEEE Geoscience and Remote Sensing Symposium, pp. 134–136. Denver, Colorado (2006)

7. Nouvel, J.F., Roques, S., du Plessis, O.: A low-cost imaging radar: DRIVE on board ONERA motorglider. In: Proceedings of IEEE Geoscience and Remote Sensing Symposium, pp. 5306–5309. Barcelona, Spain (2007)
8. Nouvel, J.F., du Plessis, O., Svedin, J., Gustafsson, A.: ONERA DRIVE project. In: Proceedings of Europe Synthetic Aperture Radar Conference, pp. 273–276. Friedrichshafen, Gemany (2008)
9. Klare, J., Weiß, M., Peters, O., Brenner, A.R., Ender, J.H.G.: ARTINO: a new high resolution 3D imaging radar system on an autonomous airborne platform. In: Proceedings of Europe Synthetic Aperture Radar Conference, pp. 381–384. Denver, Colorado (2006)
10. Klare, J., Maori, D.C., Brenner, A., Ender, J.H.G.: Image quality analysis of the vibrating sparse MIMO antenna array of the airborne 3D imaging radar ARTINO. In: Proceedings of IEEE Geoscience and Remote Sensing Symposium, pp. 5310–5314. Barcelona, Spain (2007)
11. Weiß, M., Peters, O., Ender, J.H.G.: First Flight Trials with ARTINO. In: Proceedings of Europe Synthetic Aperture Radar Conference, pp. 187–190. Friedrichshafen, Germany (2008)